Introduction to Geopolitics

Second Edition

This clear and concise introductory textbook guides students through their first engagement with geopolitics. It offers a clear framework for understanding contemporary conflicts by showing how geography provides opportunities and limits upon the actions of countries, national groups, and terrorist organizations.

This second edition is fundamentally restructured to emphasize geopolitical agency, and non-state actors. The text is fully revised, containing a brand new chapter on environmental geopolitics which includes discussion of climate change and resource conflicts. The text contains updated case studies, such as the Korean conflict, Israel–Palestine, Chechnya, and Kashmir, to emphasize the multi-faceted nature of conflict. These, along with guided exercises, help explain contemporary global power struggles, environmental geopolitics, the global military actions of the United States, the persistence of nationalist conflicts, the changing role of borders, the new geopolitics of terrorism, and peace movements. Throughout, the readers are introduced to different theoretical perspectives, including feminist contributions, as both the practice and representation of geopolitics are discussed.

Introduction to Geopolitics, Second Edition is an ideal introductory text that provides a deeper, critical understanding of current affairs, geopolitical structures, and agents. The text is extensively illustrated with diagrams, maps, photographs, and end-of-chapter further reading. Both students and general readers alike will find this book an essential stepping-stone to understanding contemporary conflicts.

Colin Flint is a Professor of Geography at the University of Illinois at Urbana-Champaign, USA.

Introduction to Geopolitics

Second Edition

Colin Flint

Routledge
Taylor & Francis Group

NEW YORK AND LONDON

First published 2011 by Routledge
2 Park Square, Milton Park, Abingdon, Oxon OX14 4RN

Simultaneously published in the USA and Canada by Routledge
711 Third Avenue, New York, NY 10017

Routledge is an imprint of the Taylor & Francis Group, an informa business

British Library Cataloguing in Publication Data
A catalogue record for this book is available from the British Library

Library of Congress Cataloging in Publication Data
Flint, Colin, 1965-
    An introduction to geopolitics / Colin Flint. — 2nd ed.
        p. cm.
    Includes bibliographical references and index.
    1. Geopolitics—Textbooks. I. Title.
    JC319.F55 2011
    320.1'2—dc23   2011024202

ISBN: 978-0-415-66772-2 (hbk)
ISBN: 978-0-415-66773-9 (pbk)
ISBN: 978-0-203-81675-2 (ebk)

Typeset in Times New Roman
by Florence Production Ltd, Stoodleigh, Devon

MIX
Paper from
responsible sources
FSC® C004839
www.fsc.org

Printed and bound in Great Britain by
TJ International Ltd, Padstow, Cornwall

CONTENTS

FIGURES

TABLES

BOXES

ACKNOWLEDGMENTS

Figures Prologue.1, 2.3, 4.2, 4.3, 4.4, 4.5, 4.6, 5.3, 6.1, 7.1, 7.2, and 7.3 were created by the talent of Muehlenhaus Map Design at www.muehlenhaus.com. I am very grateful for their skill, innovation, and vision.

Thanks to Vincent Filicetti for helping me find and collect data. I am also very grateful to Faye Leerink at Routledge for her outstanding stewardship of the book through its many stages. Also, a huge thank you to Andrew Mould for coming up with the idea of the first edition and being supportive of the second edition. I would also like to thank the reviewers of the first edition who took their time to offer thoughtful and very helpful comments and suggestions for this edition.

Finally, thanks to the students of GEOG 110 at the University of Illinois at Urbana-Champaign who are the inspiration and motivation for continuing to work on this book. Your questions, comments, looks of disbelief and puzzlement, and (occasional) supportive comments make my teaching experience an enjoyable challenge.

ABBREVIATIONS

ASEAN	Association of Southeast Asian Nations
ASSR	Autonomous Soviet Socialist Republic
BSPP	Burmese Socialist Program Party
CENTCOM	Central Command (US military)
CIA	Central Intelligence Agency
CND	Campaign for Nuclear Disarmament
CPB	Communist Party of Burma
CSTO	Collective Security Treaty Organization
DARPA	Defense Advanced Research Projects Agency
EEZ	exclusive economic zone
EU	European Union
FAO	Food and Agriculture Organization
GEOINT	Geospatial Intelligence
IMET	International Military Education and Training Program
IMF	International Monetary Fund
IPCC	Intergovernmental Panel on Climate Change
IRA	Irish Republican Army
ISI	Inter-services Intelligence
KDP	Korean Democratic Party
KNPP	Karenni National Progressive Party
KNU	Karen National Union
LORCs	Law and Order Restoration Councils
MAD	mutually assured destruction
MOOTWA	Military Operations Other Than War
NAFTA	North American Free Trade Agreement
NATO	North Atlantic Treaty Organization
NGA	National Geospatial-Intelligence Agency
NGO	non-governmental organization
NLD	National League for Democracy
NPT	Nuclear Non-proliferation Treaty
NSC	National Security Council
NSG	National System for Geospatial-Intelligence
NSS	National Security Strategy

PCCP	From Potential Conflict to Co-operation Potential
PLO	Palestine Liberation Organization
RDJTF	Rapid Deployment Joint Task Force
RFID	radio frequency identification
SCO	Shanghai Cooperation Organization
SLORC	State Law and Order Restoration Council
SSA	Shan State Army
SSA-S	Shan State Army-South
SWNCC	State-War-Navy Coordinating Committee
UN	United Nations
UNCIP	UN Commission on India and Pakistan
UNCLOS	UN Conference on Law of the Sea
UNESCO	UN Educational, Scientific and Cultural Organization
UNFCCC	United Nations Framework Convention on Climate Change
UNRWA	United Nations Relief and Works Agency
USSR	Union of Soviet Socialist Republics
WMO	World Meteorological Organization
WSF	World Social Forum
WTO	World Trade Organization

PROLOGUE

So what brings you to geopolitics? Do you see it as a way to explain the world? That would seem reasonable: yet for most of the past sixty years or so scholars, geographers in particular, distanced themselves from the topic. At the same time the desire of governments and the public for geopolitical explanations and knowledge was high, if not increasing. Why this difference between supply and demand, and how can it be addressed so that the discipline of geography is able to provide an effective framework for students, the public, and governments to understand the dynamics of world politics, or something we can call geopolitics?

What has brought many people to geopolitics, at least since the late 1800s, and continues to do so, is its apparent ability to explain in simple terms a complex and, for some, threatening and uncertain world. In offering simple explanations geopolitics can be reassuring, providing one-dimensional explanations and solutions. Such explanations are reassuring because they create the illusion of being able to know and hence to understand the world: and if we understand something it implies a relationship of control. The reassuring promises of understanding and control are reinforced by another promise of geopolitics, prediction. Geopolitical theories have always claimed an ability to tell us how the world is going to be – what and where future threats will be – and hence offer prescriptions, or policy implications (Ó Tuathail, 2006, pp. 1–2).

On the one hand, this book is designed to shatter the illusions of global understanding, prediction, control, and actionable implications by showing them to be false, dangerous, and politically motivated. But this would be simply a negative goal; a lesson in do not believe what you read. Instead, the primary intention of the book is to offer geopolitics as a framework to understand the world in its complexity as a pathway to try and explore and empathize with the diversity of political contexts and actors across the world. The emphasis is upon investigation and continual learning, knowing that we can only partially understand the situation and goals of others, rather than defining a simplified geopolitical model that is used as a tool by the powerful to proclaim what is right.

Beginning with the question of what brought you to geopolitics implies a new and purposeful engagement with an academic topic, probably as part of a university class that you have chosen, with varying degrees of freedom, to take. By the end of the book you will have learned that you have been surrounded by geopolitics continually and are always participating in it, one way or the other. The hope is that you will have learned to be critical of simple geopolitical explanations that are provided by governments, politically motivated commentators, the media, and popular culture. Also, the hope is that you will have a toolkit of your own to explore the fascinating and important topic of geopolitics. In other words, the book aims to provide you with the ability to think critically and develop your own understanding of geopolitics.

So what is geopolitics? To tease you: it is about the exercise of power. It is about geography. It is about actions. It is about how we portray, or represent, those actions. It is about how the powerful have created worlds. It is about how the weaker have resisted such efforts and, in some contexts, partially constructed their own worlds. It is about a multitude of connected actions and actors and the geographies they make, change, destroy, and maintain.

The book will explain these component parts of geopolitics and connect them. To start, the connection between geopolitics and geography will be explained, and a brief history of geopolitics offered to give you a framework for understanding the troubled history of geopolitics and the recent changes that have allowed it to reappear as an essential topic of study, but one that tries to move forward without falling into past pitfalls. The prologue ends with an outline of the purpose and framework for the book.

Geopolitics: a component of human geography

Geopolitics is a component of human geography. To understand geopolitics we must first understand what human geography is. This is easier said than done, precisely because geography is a diverse and contested discipline – in fact, the easiest, and increasingly accurate, definition is human geography is what human geographers do: accurate, but not very helpful.

Geography is a peculiar discipline in that it does not lay intellectual claim to any particular subject matter. Political scientists study politics, sociologists study society, etc. However, a university geography department is likely to house an eclectic bunch of academics studying anything from glaciers or global climate change, to globalization, urbanization, or identity politics. The shared trait is the *perspective* used to analyze the topic, and not the topic itself. Geographers examine the world through a geographic or spatial perspective, offering new insight to "sister" disciplines. For example, a political geographer may study elections or wars (as would a political scientist or scholar of international relations) but argue that full understanding is only available from a geographic perspective.

So what is a geographic perspective? In the modern history of the discipline, dominant views of what the particular perspective should be have come and gone. In the middle of the twentieth century there was an emphasis upon geography as a description and synthesis of the physical and social aspects of a region. Later, many geographers adopted a mathematical understanding of spatial relationships, such as the geographic location of cities and their interaction. Today, human geography is dominated not by one particular vision but by many theoretical perspectives, from neo-classical economics through Marxism, feminism, and into post-colonialism, and different forms of post-modernism. Furthermore, it would also be hard to think of a social or physical issue that is *not* being addressed by contemporary geography. (See Hubbard *et al.*, 2002 and Johnston and

Sidaway, 2004 to understand the history of geography and the variety of its current content; and Cox *et al.*, 2008 for a survey of contemporary political geography.)

In this book I emphasize that the geographic perspective is based upon key concepts that are shared by different theoretical frameworks. The concepts of place, space, scale, region, territory, and network will be used to explore geopolitics and, as appropriate, connect the insights made by different theories. Despite the diversity of human geography all of these concepts are used, to some degree, and provide insights into the interaction between power relations and geography. It is this interaction that underlies different approaches to geopolitics.

A diversity of geopolitical approaches

A simplified threefold classification of geopolitical approaches is used to help the reader through the history of geopolitics, the diversity of contemporary geopolitics, and the notion that what "is" geopolitics is continually contested, now more than ever. Geopolitical approaches can be classified as Classical, Critical, and Feminist.

Classical Geopolitics should not be interpreted as historic, past, and hence redundant. It is alive and well. The foundations for classical geopolitics were established in the era of European exploration and the related desire and imperative to see the world as an interconnected whole, made up of parts that were labeled or classified in a hierarchy of regions or spaces. It viewed the arena of politics as one of competition for supremacy between states. Hence, it believed that the world could be explained and understood, and as a result controlled (see Agnew, 2003 for a rich discussion of these component parts of what he calls the modern geopolitical imagination). Such understanding was the foundation for the politics of empire and colonialism; it labeled parts of the world as "barbaric" or "savage" and therefore in need of colonial control to "develop" or "civilize" their populations. Such cultural politics went hand-in-hand with a mapping of the world that cataloged it in terms of exploitable resources: gold, timber, ivory, arable land, coffee, rubber, and, not to be forgotten, cheap indigenous labor – or people.

At the end of the nineteenth century colonial competition came to a head. The supremacy of the British Empire was challenged and other countries (notably Germany, Japan, and the United States) sought to expand their colonial presence across the globe. It was in this period that the "classical" theories of people such as Sir Halford Mackinder, Alfred Thayer Mahan, and General Karl Haushofer were developed. These are discussed in more detail shortly. However, the approach of classical geopolitics lived on in the global calculations of the Cold War. Furthermore, they are prevalent today. The very act of labeling the United States' response to the terrorist attacks of September 11, 2001 the "War on Terror" was an act of geopolitics in that it identified a nebulous target that required a global military response. More precisely, the term "Axis of Evil," employed by President George W. Bush in 2002, labeled North Korea, Iraq, and Iran outside the realms of international norms and, hence, liable to military action.

In sum, classical geopolitics is a way of thinking that claims to take an objective and global perspective, but in reality has been the endeavor of elite white males in pre-dominantly, but not exclusively, Western countries with an eye to promoting a particular

Table 1 Features of classic geopolitics

Privileged position of author	White, male, elite, and Western situated knowledge
Masculine perspective	"All seeing" and "all knowing"
Labeling/classification	Territories are given value and meaning
A call to "objective" theory or history	Universal "truths" used to justify foreign policy
Simplification	A catchphrase to foster public support
State-centric	Politics of territorial state sovereignty

political agenda. Classical geopolitics has put the ideas of geographers in the service of the state, usually willingly.

In the 1990s critical geopolitics grew out of the body of thought known as post-modernism and a specific reaction by geographers to reclaim geopolitics from the state. As discussed below, in the wake of World War II geopolitics became tainted by a constructed association with the Nazi party. Geopolitics was largely practiced by government strategists rather than academics. Critical geopolitics used the tools of post-modernism to reclaim the study of geopolitics. Post-modernism is motivated by the desire to challenge statements of authority, especially those based upon science and government policy. Critical geopolitics critically engages the choice of words and the focus of policy statements, maps, essays, movies, or pretty much any media to identify what is known as the underlying discourse. Discourse is the fusion of power and authority into the content of language. For example, the common usage of "liberation" and "freedom" by US politicians and commentators through the Cold War and into the War on Terror paints pictures of moral authority and non-material gain as the basis for American foreign intervention.

Critical geopolitics used the tools of discourse analysis to re-engage the work of the past classical geopoliticians and expose their biases and political agendas. In this way it allowed for a new generation of scholars to call themselves geopoliticians; albeit critical ones who defined themselves in opposition to the classical school. Critical geopoliticians engaged current political thinkers to highlight the role of language in creating taken for granted assumptions about terrorism, Islam, the Middle East, etc. that try to generate unquestioned narratives about parts of the world, and the people that populate them, that justify military action and other foreign policy agendas. The way these understandings exist in popular culture, such as Captain America cartoon strips (Dittmer, 2010) or James Bond movies (Dodds, 2003), illustrates a point from the beginning of this Prologue, that we cannot escape geopolitics: we are exposed to it on the TV and at the movie theater as well as during politicians' speeches.

Though critical geopolitics was highly successful in bringing back the academic study of geopolitics and forcing us to think critically about what we are told about the way the world is, it too became the subject of critique. Building upon the increasing visibility and relevance of feminist thought, some pioneering scholars developed feminist geopolitics (Dowler and Sharp, 2001; Gilmartin and Kofman, 2004; Hyndman, 2004). Feminism is not simply a call to make sure that the conditions, roles, and contributions of women are given the attention they deserve, though many studies do focus on the conditions and acts of women in different geographic settings. Rather, feminism is a way of thinking that

Box 1 Geoeconomics

There has existed another approach to geopolitics, one that focuses upon the role of economics in creating geopolitical actions and theories. In the early twentieth century the Bolshevik revolutionary Lenin claimed that imperialism was an inevitable form of geopolitics given the nature of capitalism. Also, in the 1800s contrarian geopoliticians such as Kropotkin and Reclus were linking geopolitics to capitalism and suggesting alternative forms of political organization.

In the 1980s political geography engaged with a sociological theory called world-systems analysis that focused on the dynamics of global capitalism to explain two forms of geopolitics. One is the persistent differences between a few rich and powerful countries and the majority of the world who were poor and subjugated. In some periods this domination was expressed as the construction of colonies and empires while at other times, such as now, the relationship is based on unequal terms of trade and debt relations. The other is the dynamics of competition between the most powerful states and how countries have risen to be the most dominant country in the world only to lose that power to competition. The rise of British power in the nineteenth century only to lose its preeminence in the first half of the twentieth century is the most relevant case. As Britain lost its position of dominance, Germany, Japan, and the United States competed to replace it. The result was the "superpower" role of the United States and contemporary debates as to whether it is losing that position in a similar process to Britain's decline. (For more on the world-systems approach to geopolitics see Flint and Taylor, 2011.)

In the context of the United States' invasions of Iraq and Afghanistan the question of whether the need to secure oil reserves was driving US foreign policy became prominent. Such concerns led to renewed attempts to connect economics and geopolitics (Mercille, 2008), including analysis of the importance of oil (Morrissey, 2008), as well as how an economic policy that claims to be based on promoting free markets and open global trade relates to border policies and US military and political interventions (Cowen and Smith, 2009).

The connections between economics and geopolitics require prior theorization, usually from a Marxist perspective, to understand the political processes. The introductory nature of this book means that it is best to stick with geographical concepts to frame our initial exploration of what geopolitics is, allowing you to explore the geoeconomics approach subsequently.

aims to counter the simple classifications that are the underpinnings of classical geo-politics. Rather than simple, and often binary, categories, feminist geopoliticians identify the complexity of people's positions and the connectivity between people and places instead of claiming clear demarcation and difference between political spaces. The other key contribution that feminist geopoliticians make is the claim that we cannot understand the world in the top-down manner of classical geopoliticians or by simply critiquing such views, as done by critical geopolitics. What is required, feminist geopoliticians claim, is an embodied perspective; it is essential to understand what it means to be a particular individual in a particular context (i.e. a woman refugee from Darfur or a soldier on patrol in Afghanistan) to understand the way politics operates. Hence, reading and critiquing policy statements or interpreting movies is not enough; speaking to real people in real places is an empirical imperative of feminist geopolitics.

The three approaches of classical, critical, and feminist geopolitics are all alive and well and interacting with each other. The stance I take in this book is to utilize the contributions of critical geopolitics to challenge dominant classical geopolitical understandings and their imperative to categorize and create threats. In this book I also recognize that a geopolitical approach must provide understanding of the condition and actions of people in actual places, and hence I engage the ideas of feminist geopoliticians. However, I take the word "Introduction" in the book very seriously, and rather than go deeply into what can be arcane academic arguments I describe and use some key concepts to understand geopolitical actions (or practice) and the way they are represented. Before describing the organization of the book, the development of geopolitics that was briefly introduced in talking about the three geopolitical approaches will be expanded upon to give you a better sense of how and why we got here, and what the geopolitical approach of the here and now is.

A brief history of geopolitics

Geopolitics, as thought and practice, is linked to establishment of states and nation-states as the dominant political institutions. Especially, geopolitics is connected to the end of the nineteenth century – a period of increasing competition between the most powerful states – and it is the theories generated at this time that we will label "classic geopolitics." Geopolitics was initially understood as the realm of inter-state conflict, with the quiet assumption that the only states being discussed were the powerful Western countries. In other words, there was a theoretical attempt to separate geopolitics from imperialism, the dominance of powerful countries over weaker states.

Sir Halford Mackinder (1861–1947) is, perhaps, the most well known and influential of the geopoliticians who emerged at the end of the nineteenth century (Kearns, 2009). The kernel of his idea was used in justifying the nuclear policy of President Reagan, and academics and policymakers continue to discuss the merits of his "Heartland" theory. The political context from which Mackinder wrote was multi-layered. Internationally, he was concerned about the relative decline in Great Britain's power as it faced the challenge of Germany. Within Britain, his conservatism was appalled by the destruction of traditional agricultural and aristocratic lifestyles in the wake of industrialization, especially the rise of an organized working class that made claims for social change. His goal was to maintain both Britain's power and its landed gentry through a strong imperial bloc that could resist challengers while maintaining wealth and the aristocratic social structure.

Influenced by the work of Alfred Thayer Mahan (1840–1914), Mackinder saw global politics as a "closed system" – meaning that the actions of different countries were necessarily interconnected, and that the major axis of conflict was between land- and

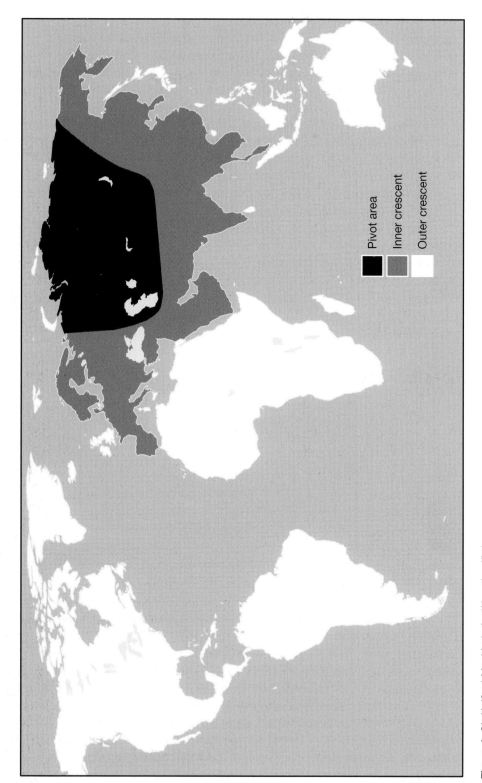

Figure 1 Sir Halford Mackinder's "Heartland" theory.

sea-powers. He defined the geography and history of land-power by designating, in 1904, the core of Eurasia as the Pivot Area, which in 1919 he renamed the Heartland. This area was called the Pivot Area because, in his Eurocentric gaze, the history of the world pivoted around the sequence of invasions out of this region into the surrounding areas that were more oriented to the sea. In the past, Mackinder believed, sea-powers had maintained an advantage, but with the introduction of railways, he reasoned, the advantage had switched to land-powers; especially if one country could dominate and organize the inaccessible Heartland zone. Hence Mackinder's famous dictum, or, in contemporary language "tweet":

> Who rules East Europe commands the Heartland
> Who rules the Heartland commands the World Island
> Who rules the World Island commands the World.

The World Island was Mackinder's term for the combined Eurasian and African landmasses.

Mackinder's twin goals were to maintain British global preeminence in the face of challenge from Germany, the country most likely to "rule" Eastern Europe and, in the process, resist changes to British society. After initially discounting the role of the United States, in 1943 he proposed a Midland Ocean Alliance with the US to counter a possible alliance between Germany and the Soviet Union. Following his identification of the Heartland, roughly representing the territorial core of the Soviet Union, plus his emphasis on alliance, Mackinder was the intellectual basis for Cold War strategists and proponents of the North Atlantic Treaty Organization.

Mackinder's contribution is also a good illustration of two prevalent features of "classic" geopolitics. First, he used a limited and dubious Western-centric "theory" of history to claim an objective, neutral, and informed intellectual basis for what is in fact a very biased or "situated" view with the aim of advocating and justifying the policy of one particular country. Plus, he disseminated a catchy phrase or saying to influence policy. Second, Mackinder's career is one of many examples of the cross-over between academic or "formal" geopolitics and state policy or "practical" geopolitics: he was a successful academic, founding the Oxford School of Geography in 1899 and serving as director of the London School of Economics between 1903 and 1908.

Alfred Thayer Mahan (1840–1914) also walked in academic and policy circles. He rose to the rank of admiral in the US Navy and was president, at different times, of both the Newport War College and the Naval War College. His two books *Influence of Seapower upon History* (1890) and *The Interest of America in Seapower* (1897) were important influences upon Presidents McKinley and Theodore Roosevelt, as well as the German Kaiser Wilhelm II. Mahan made a historical distinction between land- and sea-powers that was to influence geopolitical thinkers through the Cold War. He believed that great powers were those countries whose insularity, coupled with an easily defensible coastline, provided a secure base from which, with the aid of a network of land bases across the world, sea-power could be developed and national and global power attained and enhanced. In addition, Mahan advocated an alliance with Britain to counterbalance

Eurasian land-powers. His influence upon Mackinder is clear, but Mahan's goal was to increase US global influence and reach, while avoiding conflict with the dominant British Navy.

The United States was not the only country which was eyeing Great Britain's supremacy. In Germany, politicians and intellectuals viewed Britain as an arrogant nation that had no "divine right" to its global power. In the words of Chancellor Bismarck, Germany deserved its "place in the sun." "German" geopolitics was defined by the work of two key individuals: Friedrich Ratzel (1844–1904) and Rudolf Kjellen (1864–1922). Similar to his English counterpart Mackinder, Ratzel was instrumental in establishing geography as an academic discipline. Furthermore, his *Politische Geographie* (1897) and his paper "Laws of the Spatial Growth of States" laid the foundations for *geopolitik*. However, it was the Swedish academic and parliamentarian Kjellen who developed Ratzel's idea and refined an organic view of the state. Following Ratzel's zoological notions, Kjellen propagated the idea that states were dynamic entities that "naturally" grew with greater strength. The engine for growth was "culture." The more vigorous and "advanced" the culture the more right it had to expand its "domain" or control more territory. Just as a strong pack of wolves could claim hunting grounds of a neighboring but weaker pack, the organic theory of the state asserted that it was more efficient and "natural" for advanced cultures to expand into the territory of lesser cultures. Of course, given the existing idea that cultures were contained within countries or states, this meant that borders were movable or expandable. The catchphrase for these ideas was Ratzel's *Lebensraum*, or living space, meaning that "superior," in the eye of the beholder, cultures deserved more territory as they would use the land in a better way. In practice, the ideas of Ratzel and Kjellen were aimed at increasing the size of the German state eastwards to create a large state that the "advanced" German culture warranted, in their minds at the expense of the Slavs who were deemed culturally inferior.

The German process illustrates a key feature of classic geopolitics: the classification of the earth and its peoples into a hierarchy that then justifies political actions such as empire, war, alliance, or neglect. This process of social classification operates in parallel with a regionalization of the world into good/bad, safe/dangerous, valuable/unimportant, peaceful/conflictual zones. Dubious "theories" of the history of the world and how it changes are used to "see" the dynamics of geopolitics as if from an objective position "above" the fray: Haraway's (1998) God's eye view. Of course, we should note the influential positions of these geopoliticians. Geopolitical theorists are far from being neutral, objective, and uninterested.

Before we move on to the Cold War period, we should briefly return to the German school of geopolitics to make a couple more points about classic geopolitics in general. As Adolf Hitler and the Nazi party began to rise to power in the 1920s, General Karl Haushofer (1869–1946) began to disseminate geopolitical ideas to the German public through the means of a magazine/journal entitled *Zeitschrift für Geopolitik* (Journal of Geopolitics) and a weekly radio show. Haushofer was skillful in creating a geopolitical vision that unified two competing political camps in inter-war Germany: the landed aristocrats, who wanted to expand the borders of Germany eastwards towards Russia and the owners of new industries such as chemicals and engineering who desired the establishment of German colonies outside of Europe to gain access to raw materials and markets. This idea came together in his definition of pan-regions (large multi-latitude regions that were dominated by a particular "core" power). In this scenario, the US dominated the Americas and Germany dominated Eurasia while Britain controlled Africa. Haushofer's vision allowed for both territorial growth and colonial acquisition for Germany, without initiating conflict with Britain.

Haushofer blended a policy, and made the German public aware of foreign policy debates, that ran parallel with Hitler's surge in popularity and his vision of a "strong" Germany. However, Haushofer was not Hitler's "philosopher of Nazism" as *Life* magazine famously declared in 1939 (Ó Tuathail, 1996, p. 115). In fact, there was a significant difference between the views of Haushofer – with his emphasis on geographic or spatial relationships – and Hitler, whose racist view of the world shaped his geopolitical strategy. But the point is that Haushofer did use Hitler's surge to power as a means of advancing his own career. Haushofer's tragic tale (he ultimately committed suicide following questioning by the US after the war regarding his role as a war criminal) has resonated throughout the community of political geographers ever since. Equating "geopolitics" with the Nazis tainted the sub-discipline of political geography and it practically disappeared as a field of academic inquiry immediately after World War II.

However, there is another lesson to take from Nazi geopolitics too – and that is how it continues to be portrayed by academics. Many recent studies have contextualized and examined the content of Nazi geopolitics in depth. Not to apologize for their connection

---

Box 2 Geodeterminism

> Geopolitics is the science of the conditioning of political processes by the earth. It is based on the broad foundation of geography, especially political geography, as the science of political space organisms and their structure. The essence of regions as comprehended from the geographical point of view provides the framework for geopolitics within which the course of political processes must proceed if they are to succeed in the long term. Though political leadership will occasionally reach beyond this frame, the earth dependency will always eventually exert its determining influence.
>
> (Haushofer *et al.*, 1928, p. 27, quoted in O'Loughlin, 1994, pp. 112–13)

The quote from General Haushofer offers an example of the "geodeterminism" of classic geopolitics, or the way in which political actions are determined, as if inevitably, by geographic location or the environment. Such an approach can be used to justify foreign policy as it removes blame from decision-makers and places the onus on the geographic situation. In other words, if states are organisms then Germany's twentieth-century conflicts with its neighbors are represented as the outcome of "natural laws" and not decisions made by its rulers.

to Hitler but to place the development of their theories within the contexts of global politics and the development of academic thought. The research shows there were indeed differences between their theories and Hitler's vision. Also, another outcome of this work is to show that Mackinder shared some of the academic baggage of the German geopoliticians. The predominance of biological analogies in social science at the end of the nineteenth century and the beginning of the twentieth meant that Mackinder and the German school were influenced by ideas that equated society with a dynamic organism. The key difference was that Mackinder was writing from, and for, a position of British naval strength, while the Germans were trying to challenge that power through continental alliances and conflicts with a wary and envious eye on British sea-power.

Post World War II there existed an interesting irony: the vilification of "geopolitics" as a Nazi enterprise resulted in its virtual disappearance from the academic scene. On the other hand, as the United States began to develop its role as a post-war world power it generated geopolitical strategic views that guided and justified its actions. Prior to World War II, Isaiah Bowman (1878–1956), one-time President of the Association of American Geographers, offered a pragmatic approach to the US's global role, and was a key consultant to the government, most notably at the Treaty of Versailles negotiations at the end of World War I. Nicholas Spykman (1893–1943), a professor of International Relations at Yale University, noted the US's rise to power and argued that it now needed to practice balance of power diplomacy, as the European powers had traditionally done. Similar to previous geopoliticians, Spykman offered a grandiose division of the world: the Old World consisting of the Eurasian continent, Africa and Australia and the New World of the Americas. The US dominated the latter sphere, while the old world, traditionally fragmented between powers, could, if united, challenge the United States. Spykman proposed an active, non-isolationist US foreign policy to construct and maintain a balance of power in the "Old World" in order to prevent a challenge to the United States. Spykman identified the "Rimland," following Mackinder's "inner crescent," as the key geopolitical arena. In contrast to the calls for greater global intervention, Major Alexander P. De Seversky (1894–1974) proposed a more isolationist and defensive stance. His theory is notable for its emphasis upon the polar regions as a new zone of conflict, using maps with a polar projection to show the geographical proximity of the US and the Soviet Union, and the importance of air-power.

Increasingly, US geopolitical views took the form of government policy statements that, in the absence of academic endeavors, assumed the status of "theories," and hence gained an authority as if they were objective "truths." First came George Kennan's (1904–2005) call for containment, then NSC-68's call for a global conflict against communism, supported by the dubious "domino theory." These geostrategic policy statements will be discussed in greater depth in Chapter 2. In the relative absence of academic engagement with the topic, geopolitical theories were constructed within policy circles, and, despite the global role of the US, a limited perspective remained. George Kennan, for example, is identified as a "man of the North [of the globe]" who identified the Third World as "a foreign space, wholly lacking in allure and best left to its own, no doubt, tragic fate." Kennan, in the tradition of his academic predecessors, was also eager to classify the world into regions with political meaning, defining a maritime trading world (the West) and a despotic xenophobic East.

Box 3 "Western-centrism" and "geopolitical traditions"

Critical engagement with the history of geopolitics has focused on the scholars and practitioners in European countries and the United States. This is unsurprising and, to some extent, justifiable given the role of Mackinder, Ratzel, and Haushofer in creating and promoting modern geopolitics. However, the form of geopolitics these figures, along with Mahan, created was deemed not only applicable but a strategic necessity in many other countries. Notably, Japan, as part of the construction of an Asian empire in the late nineteenth and early twentieth century, created its own geopolitical framework. Specifically, the way Manchuria was constructed as a geopolitical region to justify Japanese imperial expansion was theorized (Narangoa, 2004).

The key features of classical geopolitics framed the content of theories created in non-Western contexts, but the particular circumstances of those contexts produced nuances and different emphases. The idea of "geopolitical traditions" (Dodds and Atkinson, 2000) is a useful way to explore the combination of consistent dominant themes and specifics of historical-geographical context in geopolitical thinking. A collection of essays by Dodds and Atkinson (2000) was a significant contribution in forcing recognition of non-Western forms of geopolitics. The second edition of the *Geopolitical Reader* highlighted a more diverse range of statements made from within the Soviet Union. The particular forms of Brazilian and South African geopolitics have also been noted.

Increasingly, non-Western geopolitics, both contemporary and historic, is being illuminated by researchers. Though the "founding fathers" of modern geopolitics may always give a Western-centric bias to the study of the history of geopolitics this bias is being diluted to some degree. Furthermore, the importance of ongoing geopolitics in South and East Asia, the Middle East, and Africa will mean that contemporary analysis will, to some degree, ensure a more global coverage.

Perhaps, in hindsight, the lack of policy-oriented geopolitical work in the academic world provided room for the critical understandings of geopolitics that now dominate the field. With the exception of Saul Cohen's (1963) attempt to provide an informed regionalization of the world to counter the blanket and ageographical claims of NSC-68, geographers were largely silent about the grand strategy of inter-state politics. However, with the publication of György Konrád's *Antipolitics* (1984), in accordance with other theoretical developments in social science thinking and public dissent over the nuclear policies of Ronald Reagan, geographers found a voice that produced the field of "Critical Geopolitics" as well as broader systemic theories about international politics (see box Prologue.1). Both of these approaches, though very different in their content and theoretical frameworks, offered critical analysis of policy, rather than being a support for government policy.

Though it is hard to summarize the diversity of these approaches, there is one important commonality: the study of geopolitics is no longer state-centric. Geopolitical knowledge is now understood and critiqued as being "situated knowledge." Though this observation has been used to claim the relevance of the perspectives and actions of contemporary marginalized groups, it may still be used to consider the thoughts of the theoreticians we have just discussed, whose concern was geopolitical states*man*ship. In other words, geopolitical theoreticians constructed their frameworks within particular political contexts and within particular academic debates that were influential at the time, the latter sometimes called paradigms.

Current geographical analysis aims to contextualize the actions of particular countries or states within their historical and geographical settings. For example, the decisions made by a particular government are understood through the current situation in the world as a whole. It is this approach that guides most of the content of this book. Critical geopolitics "unpacked" the state by illustrating that it is impossible to separate "domestic" and "foreign" spheres, that non-state actors – such as multi-national companies, non-governmental organizations and a variety of protest groups and movements for the rights of indigenous peoples, minorities, women, and calling for fair trade, the protection of the environment, etc. – play a key role in global politics.

The bottom line: geopolitics is no longer exclusively the preserve of a privileged male elite who used the authority of their academic position to frame policy for a particular country. Though these publications still exist, most academics who say they study geopolitics are describing the situation of those who are marginalized, and advocating a change in their situation. Study of the state is often critical, but it is just one component of a complicated world – rather than a political unit with the freedom to act as the theory suggests it should in a simplified and understandable world.

This brief history of geopolitics is intended to introduce you to the role and content of "classic" geopolitics and the growth of alternative geopolitical frameworks. Two words of caution. First, and as noted, this history is Eurocentric. I urge the reader to use the *Dictionary of Geopolitics* (O'Loughlin, 1994) to see how thought in countries such as Japan and Brazil reflects and differs from those discussed above. Japan, for example, had its own debate about the merits of the German school of geopolitics, with the ideas of Ratzel and Kjellen being popular amongst Tokyo journalists but less so within academic circles. Second, do not be fooled by the prevalence of "critical geopolitics" in the academy. Bookstores are continually replenished by volumes purporting to "know" everything about "Islam," "terrorists," and a variety of imminent or "coming" wars. Some of these volumes are quite academic, and others more popular. They all share the arrogance of claiming to be able to predict the future and, hence, are assured about what policies should be adopted. "Classic" geopolitics lives, but now it must contend with an increasingly vigorous and confident "critical geopolitics." In other words, geopolitics is itself a venue and practice of politics.

Organization of the book

Chapter 1 begins the book with the introduction of a simplified model of global geopolitics. It ends with a discussion of the complexity, or "messiness," of geopolitical conflicts given the multiplicity of structures and the multiple identities and roles of agents. The text assumes no familiarity with geopolitical terms and no prior knowledge of conflicts, past or present. As you progress through the book, try to make your own understanding of geopolitics more sophisticated by exploring how the different structures and agents introduced in successive chapters interact with one another. Also, be engaged with quality newspaper and other media reports of current events. Use the text and the current events (i) to identify the separate structures and agents and then (ii) to see how they are related to each other. In other words, allow yourself to explore the complexity of geopolitics as you work through the book and become familiar with a growing number of structures and agents.

Within the overarching idea of structure and agency the book will be organized in the following way. Chapter 2 focuses attention upon the scale of countries, especially the choices and constraints they face as geopolitical agents. The foreign policy that negotiates these choices and constraints is called a geopolitical code. Chapter 3 remains with the topic of geopolitical codes, but shows the importance of how they are justified or represented. The representation of geopolitical codes is important for a country, in order for its actions or agency to be supported rather than contested.

Chapter 4 addresses geopolitical agents that construct and contest the state scale, as we formalize our understanding of countries by introducing them as states, and discussing the related concepts of nation, nationalism, and nation-states. The ideology of nationalism and the geopolitics of separatism are topics discussed in this chapter. Nationalism is a collective identity creating the assumption of community at the national scale and the correspondence of that identity with the spatial organization of society into nation-states. The ideological maintenance of states through nationalism is complemented through their territorial expression. Chapter 5 addresses the geopolitics of territory, boundaries, and boundary disputes as the means of defining the geographic expression of states.

From focusing upon the geopolitical agency and structural context of states, Chapter 6 introduces another geographical expression of power, networks. The expressions "global terrorism" and "globalization" are common contemporary understandings that politics involves the movement or flow of things across boundaries and into the jurisdiction of states. These flows are both legal and illegal. The flows are facilitated by networks, whether a terrorist or criminal network, on the one hand, or the network of global finance that switches huge amounts of money from financial market to financial market across the globe. In Chapter 6 we will focus upon the topics of terrorism and social movements.

Chapter 7 concentrates upon the global scale, and provides a way of thinking of a dynamic global geopolitical context, the structure within which state and non-state actors must operate. We do this through a critical engagement with a historical model of the rise and fall of great powers, more specifically George Modelski's cycles of world leadership. Chapter 8 brings together the concepts of geopolitical codes, structure and agency, nation-states, networks, and state and non-state actors through a focus upon the increasingly important topic of environmental geopolitics. The two issues of global climate change and resource conflicts are used as examples.

Chapter 9, the final chapter, summarizes the identification of geopolitical structures and agents, but complicates the picture by showing how contemporary conflicts are usually a combination of the structures and agents that have been treated separately in the preceding chapters. The book concludes by challenging you to continue to explore the role of geography in causing, facilitating, and concluding geopolitical conflicts: both those ongoing and those yet to come.

1

A FRAMEWORK FOR UNDERSTANDING GEOPOLITICS

In this chapter we will:

- Define the key concepts of place, scale, region, territory, and network
- Introduce the concept of structure and agency
- Show how place, scale, region, territory, network, and structure and agency will be used to understand geopolitics
- Define geopolitics
- Consider what is "power"
- Provide examples of these concepts
- Use our own experiences and knowledge to understand and investigate these concepts

Geopolitics is part of human geography. We can use the perspective of human geography to understand how politics, especially international politics, and geography are related. This chapter introduces some fundamental concepts of human geography that will be used throughout the book to explain what geopolitics is and to understand contemporary conflicts and issues. The concepts of place, space, scale, region, network, and territory will be introduced and used to define geopolitics. The final part of the chapter introduces two other important concepts we will use to understand geopolitics: the interaction between structure and agency, and power.

By the end of the chapter you will have a better sense not only of the human geography perspective but also of how it can be used as a framework for understanding and explaining geopolitics. You should also come away with two important overarching ideas. The first is contestation: what places, regions, territory are and who belongs are continually politicized in geopolitics of inclusion and exclusion. In other words, geography is always political and in some instances the politics is violent. The second idea is context: the political events that are reported in the media happen within different geographical and historical settings which partially define what happens and what possibilities for peace and resolution exist.

Geography and politics

Human geography may be defined as: the *systematic study of what makes places unique and the connections and interactions between places* (Knox and Marston, 1998, p. 3). In this definition human geographers are seen to focus upon the study of particular neighborhoods, towns, cities, or countries (the meaning of place being broad here). In other words, geographers are viewed as people who study the specifics of the world – not just where Pyongyang is, but what its characteristics are. "Characteristics" may include weather patterns, physical setting, the shape of a city, the pattern of housing, or the transport system. Political geographers are especially interested, amongst other things, in topics such as how the city of Pyongyang, for example, is organized to allow for political control in a totalitarian country.

However, places (whether neighborhoods or countries) are not viewed as isolated units that can only be understood through what happens within them. The first definition also highlights the need to understand places in relation to the rest of the world. Are they magnets of in-migration or sources of out-migration? Are investors of global capital seeking to put their money in a particular place, or are jobs being relocated to other parts of the world? Is a place a site for drug production, such as areas of Afghanistan, or the venue for illegal drug use – such as suburban areas of the United States or Europe? Understanding a place requires analyzing how its uniqueness is produced through a combination of physical, social, economic, and political attributes – and how those attributes are partially a product of connections to other places, near and far.

A further and complementary definition of human geography is: *the examination of the spatial organization of human activity* (Knox and Marston, 1998, p. 2). In this definition space is emphasized rather than place. The term space is more abstract than place. It gives greater weight to functional issues such as the control of territory, an inventory of objects (towns or nuclear power stations, for example) within particular areas, or hierarchies and distances between objects. For example, a spatial analysis of drug production and consumption would concentrate on quantifying and mapping the flows of the drug trade, while an emphasis of place would integrate many influences to understand why drugs are grown in some places and consumed in others.

The economic, political, and social relationships that we enjoy and suffer are mediated by different roles for different spaces. Two banal examples: if you are going to throw a huge and rowdy party, don't do it in the library; as a student, when entering a university lecture hall sit in one of the rows of seats rather than standing behind the lecturer's podium. The banality of these examples only goes to show that our understanding of how society is spatially organized is so embedded within our perceptions that we act within subconscious geographical imaginations. In addition, these two examples also show that the spatial organization of a society reflects its politics, or relationships of power. Standing behind the lecturer's podium would be more than an invasion of her "personal space" but a challenge to her authority: It would challenge the status quo of student–lecturer power relationships by disrupting the established spatial organization of the classroom.

Compare the maps of Africa in Figure 1.1. The maps display two spatial organizations of power relations. The large map illustrates the spaces of independent countries (or states)

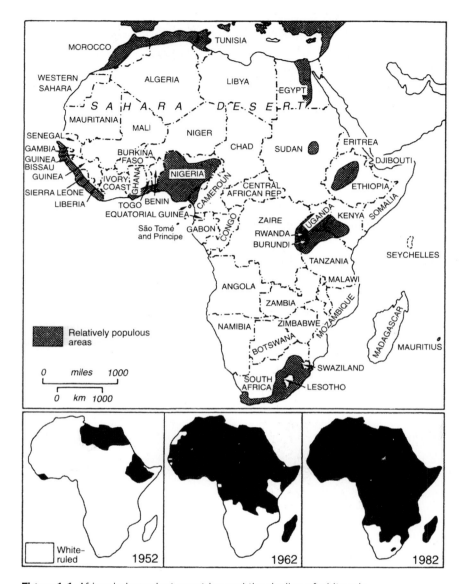

Figure 1.1 Africa: independent countries and the decline of white rule.

that were created after the decline of the colonial control imposed by European powers in the nineteenth century. External powers defined parts of Africa as "theirs," and so allowed them to subjugate the native populations for perceived economic benefit free of violent and costly competition with other European countries. These spaces were a product of two sets of power relations: the ability of European countries to dominate African nations and the relative power parity amongst the European countries. The map of countries is a different spatial organization of power in Africa, the post-colonial establishment of independent African countries. This new spatial organization of power reflects a relative decrease in the power of the European countries to dominate Africa, though a hierarchy of power remains. However, focusing on the spaces of independent countries across the continent obscures other power relations, especially those of gender, race, and class relations *within* the countries. As we shall see, the scale at which we make our observations highlights some political relations and obscures others. The three smaller maps depict the struggle of Africans to end white rule of their countries after the end of the colonial period. It shows the racialized spaces of political control, the areas of Africa in which the descendants of European settlers were able to maintain control, and how these spaces of white control have now shrunk to nothing, given the end of apartheid in South Africa.

Places and politics

First, let's focus more closely on defining what we mean by place. From our earlier definition of human geography we know that places are unique and interdependent. In addition, *places are the settings of people's everyday lives* (Knox and Marston, 1998, p. 3). In other words, people's daily experiences, whether it be dodging mortar rounds in Baghdad or enjoying the wealthy trappings of up-scale housing in the gentrified London docklands, are a reflection of where they live. Life-chances are still very much determined by where one is born and grows up. Table 1.1 shows infant mortality rates across the globe, but also how this varies within countries; the United States is used as an example. The life expectancy in the state of Delaware, for example, is similar to the national average in Chile. What we may do, what we are aware of, what we think and "know" are a function of where we live. Places are the sites of employment, education, and conversation. Since places are unique they will produce a mosaic of experiences and understandings.

To better understand how geographers think about place, we will use two different authors. First, John Agnew's (1987) definition suggests that places are the combination of three related aspects: location, locale, and sense of place.

Table 1.1 The geography of infant mortality rates (per 1,000 live births)

Variation across countries 2009		Variation within the United States 2007	
United States	7	New Hampshire	5.4
Austria	3	Utah	5.1
Greece	3	California	5.4
Romania	10	Arizona	6.9
Chile	7	Georgia	8
Colombia	16	Alabama	9.9
Singapore	2	Mississippi	10
China	17	District of Columbia	13.1
Vietnam	20	Delaware	7.5
Pakistan	71		
Egypt	18		

Activity

Stop reading for a minute and write down four or five features of your hometown that make it distinctive. We will be referring back to these features and developing them as we go through this chapter.

My example: I grew up near Dover, Great Britain. It is a major ferry port connecting the British Isles to the European continent. Surrounding the town was a scattering of coal mines that were closed down after the miners' strike of 1984. The landscape of Dover is dominated by Dover castle situated on the cliffs; the castle keep dates from Norman times, and within the grounds are the ruins of a Roman lighthouse. The opening of the Channel Tunnel has threatened the profitability of the cross-channel ferries. In the past few years the town has experienced the presence of refugees from eastern and southeastern Europe. Though I am emotionally connected to Dover, I do find it a bit drab.

Location is the role a place plays in the world, or its function. The key industries and sources of employment within a place are a good measure of location – whether it is a steel mill, a coal mine, a military base, or a tourist resort. Of course, these are simplistic examples, and usually places will be a combination of different functions – perhaps complementing each other or existing together uneasily. Dover, Great Britain is a good example, where its function as a ferry port promoted another function, the point of entry for refugees.

Locale refers to the institutions that organize activity, politics, and identity in a place. People operate as parts of groups: families, schools, workplaces, communities of worship, labor unions, political parties, militias, parent–teacher organizations, sports clubs, etc. In combination, these institutions form the social life of a particular place. The wave of social protest in Arab countries in 2011 was a battle over locale, or the types of political institutions within the countries. The protestors wanted to overthrow non-democratic and despotic forms of government and replace them with something more democratic. Interestingly, the army was often seen as an institution that could help in a progressive move toward democracy. Within the context of Arab politics the army has a different relationship with democracy than in established Western democracies. Underlying the protests in some of the countries were claims that political institutions favored the majority Sunni populations and marginalized Shi'ite communities. Hence the politics of locale must be related to the politics of identity, and the final component of Agnew's definition of place.

The third aspect of place is *sense of place*. People's identity is a function of membership in a number of collective identities: gender, race, social class, profession, nationality, and, last but not least, place. Sense of place is a collective identity tied to a particular place, perhaps best thought of as the unique "character" of a place. People are guided in their actions by particular identities that say who they are and what they can and cannot, should and should not, do. Belonging to a particular ethnic group socializes people into particular expectations and life chances. Part of one's sense of "belonging" is attachment to place, which can translate into visions of what a place should be "like": Notably, who "belongs"

Box 1.1 Place and Palestinian identity

"The man who entered the room was visibly distraught. Wasting no time on pleasantries, he threw himself down in a chair and announced that the soldiers had gone berserk. This was in early 1994, before the Israeli pullback. Just before midnight, the man said, twenty or thirty people in the al-Boureij refugee camp had been forced out of their homes. Some of them hadn't even had a chance to put on their shoes; others complained that the soldiers had kicked and hit them. They were led to the UNRWA school [United Nations Relief and Works Agency for Palestine Refugees in the Near East] and ordered to pick up some garbage and rocks that had been strewed in the yard. Furious, the man said that someone was made to write slogans in Arabic on the wall. 'Life is like a cucumber,' was the worst of them. 'One day in your hand, the next day up your ass.'

"During the four years of the intifada, Kafarna [the man who entered the room] had witnessed countless violent clashes and far greater indignities than those he recounted that day, but he never managed to come to terms with any of it; he was simply unable to swallow the insult" (Hass, 2000, p. 31).

In this quote, note not only the evocation of the Gaza Strip as a place, but how Kafarna's individual identity and politics are inseparable from his place-specific experiences.

and who doesn't. A harmless example is the urban myth underlying the self-proclaimed moniker of the "Dover Sharks." The name is dubiously derived from stories of piracy from centuries ago when the locals would set lights to confuse the Channel shipping and induce wrecks. They would then wait for and loot the cargo that washed ashore. The name "Dover Sharks" is claimed to derive from the habit of freeing rings from the bloated corpses by biting off their fingers! More disturbing is the way that Dover residents reacted to the refugee "outsiders" as undeserving recipients of local government services that should only be available to residents of Dover. To relate place identity to contemporary conflict, the quote in Box 1.1 is an observation of a Palestinian man whose political beliefs are clearly tied to his attachment to place.

Using a sense of place to construct an identity politics of insiders and outsiders (Cresswell, 1996) is evident in the language of hate politics in the US when those deemed "not to belong" enter places with dominant or established "identities" and "traditions." Such politics are frequently racial, but homosexuals are regularly targeted too. Gentrified neighborhoods that are known to have large gay populations are often associated with anti-gay hate crimes that are spurred not solely by homophobia, but also by indignation over how the place has changed from an understanding of its "traditional" form. For example, particular visions of place are evident in anti-gay hate crimes in the Victorian Village neighborhood of Columbus, Ohio:

When you have somebody who, say, has a rainbow flag on their house and then you drive by and you see that it's been torched, it's not just a crime against the people living in that house. That is sending a message to the entire community the same way that having somebody put a cross on somebody else's front lawn sends a message to the entire community.

(Executive Director, Buckeye Region Anti-Violence Organization, May 29, 2001. Quoted in Sumartojo, 2004, p. 100)

In another example, the murder of Dutch filmmaker Theo van Gogh by Islamic radicals in the Netherlands in November 2004 sparked a debate about the political future of a country proud of its record of tolerance. One Amsterdam newspaper, the *Algemeen Dagblad*, claimed that anti-Muslim graffiti suddenly appeared everywhere, and attacks on mosques and other Islamic buildings erupted. However, the Amsterdam newspaper *Trouw* argued that such reports were overblown and reported the Prime Minister Jan-Peter Balkenende as saying: "We must not allow ourselves to be swept away in a maelstrom of violence . . . Free expression of opinion, freedom of religion, and other basic rights are the foundation stones of our state . . . all moderates will join together in the fight against the common enemy: extremism" (all quotes from *The Week*, November 26, 2004, p. 15).

Politics of inclusion and exclusion are dominated by issues of immigration as well as the ethnic and religious character of a nation (see Chapter 4). For example, some extremist groups claimed that President Obama was a Muslim, despite his constant and public affirmation of his Christian faith, to portray him as alien and somehow dangerous to what is implicitly proclaimed to be a white Christian United States. As a general conclusion, the function of a place, which social groups have control of the institutions within a place, and the identity of a place, are contested. Racialist and homophobic groups are extreme examples of the politics of place.

Activity

Refer back to the features of your hometown you identified previously. Classify these features using Agnew's three aspects of place. If one of the aspects is not included, think of a feature of your hometown that would fit.

Can you find examples in your local and national newspapers of how the location, locale, and sense of place of your hometown have been or still are contested?

An alternative view of society emphasizes cosmopolitanism – attachment to no particular place. Globalization is seen by some to have created a class of "global citizens" who travel across the globe on business and political trips, or even for leisure. Focusing upon this relatively small group of people should not detract from the fact that it is the socially privileged (in terms of wealth, race, and gender) who have the status and ability to travel easily from place to place and feel at "home" wherever they are (Massey, 1994, p. 149).

ıs more worthy of our attention is the role of diasporas – networks of migrants ˙sh connections between places across the globe. A good example is the Chinese ᵧ in Vancouver, Canada that has facilitated massive amounts of investment by .ᵢese capitalists (large and small) in the real estate economy of British Columbia. Diasporas illustrate how a person can be attached to a number of places, though perhaps this geography may mean they are not completely "at home" anywhere.

The second author we will discuss is Doreen Massey. Her definition of place complements John Agnew's. For Massey, places are networks of social relations "which have over time been constructed, laid down, interacted with one another, decayed and renewed. Some of these relations will be, as it were, contained within the place; others will stretch beyond it, tying any particular locality into wider relations and processes in which other places are implicated too" (Massey, 1994, p. 120).

Massey's definition gives us three extra points to consider about places. First, they are the products of human activity, or in social science parlance they are "socially constructed." The functions of a place, the institutions within it, and its character stem from what people do. The Arab democratization protests of 2011 were a product of individual actions, the groups they formed, and the construction of a progressive resistance identity within national and religious histories. When referring to "social relations," Massey identifies social hierarchies formed within the workplace, between racial and religious groups, and also the pervasive influence of gender upon normative expectations.

Second, places are dynamic or they change over time. What people, do, want, and think changes over time and such aspirations are translated into projects that make and remake places. The Arab countries are now different from the time I wrote these words. Social movements and political groups that support different options for change, or oppose any change, are continually interacting to make the places and politics fluid. The identity of the Arab protestors and their goals have changed over the course of the protests. In another example, the landscape of contemporary Moscow is made up of layers from the Soviet past and its celebration of Communism and contemporary signs of consumer capitalism (see Figure 1.2).

Third, and related to our first definition of human geography, places can only be understood fully through their interactions with other places. Returning once more to the 2011 Arab democratization movements, the role of the United States and other "outside" countries was very influential in defining support, or lack of it, for rulers trying to cling to power. Outside countries also tried to define what the process of change would look like. In February 2011 British Prime Minister David Cameron visited Egypt, declaring his country was a "candid friend," and urged the then interim military rulers to stand by their promise of organizing free and fair elections quickly. Also, the very social movements themselves were connected by social media and influenced each other's actions: the initial act of self-immolation in Tunisia was a single event in one place that catalyzed protest and political action in many others.

Massey's emphasis upon the dynamism of place and Agnew's recognition of institutional politics and sense of place illustrate the central role contest or conflict plays in defining places. Let's go back to our earlier banal examples of the interaction between space and politics. Partying in the library would be an act that challenged the norms and

Figure 1.2 Lenin Statue, Moscow.

Activity

How have the location, locale, and sense of place of your hometown changed over time?

How is your hometown connected to other places?

How does your hometown's past history influence its present and future?

rules of a particular place: a political act to change the function, meaning, and ambience of the library.

In a more significant example, Okinawa, Japan has been dominated by US military bases since the end of World War II. In Agnew's (2003) terminology, Okinawa's location is defined by its geostrategic military role for the US. An article in the *New York Times* of September 13, 2004 highlighted the aftermath of a military helicopter crash in which the Japanese police were not allowed to investigate the crash-site. Through Agnew's lens, different institutions were in a jurisdictional contest over the territory of Okinawa: who is in control the US military or the Japanese police? Such contestation led to local protest, and expressions of self-identity reflected in a local high school teacher's sentiments: "At that time I felt Okinawa is really occupied by the US, that it is not part of Japan."

The role of Okinawa, the power of the local police in relation to the US military, and the identity of the place are contested, especially inflamed by incidents such as the helicopter crash. Places contain many different institutions, and collective identity is usually multi-dimensional. So, places are sites of multiple conflicts. In the case of Okinawa, the conflict over the US presence is connected to the situation of the island within Japan. The *New York Times* article went on to quote the female teacher, claiming: "Tokyo doesn't care . . . I feel a gap between Tokyo and here." In other words, the contestation of what Okinawa is and will become is a combination of a regional identity rejecting the authority of the Japanese government and an assertion of local authority over the American presence.

What other contests are likely in play here but are not mentioned in this article? Some hints include the gender of the teacher and the mention in the article of demonstrations a decade earlier after the rape of a twelve-year-old schoolgirl by three American servicemen.

The protests in Okinawa illustrate many points we have covered: Agnew's three aspects of place should be understood as connected rather than separate entities; the nature of a place is a function of its connections to the outside world; places are contested; and the contestation produces dynamism – places change. The Okinawa example also introduces us to another important geographic concept, scale. It is impossible to interpret the actions of the Okinawan high school teacher without taking into consideration the island, its position within Japan, and the US's global military presence. It is to geographic scale that we now turn.

Activity

Now is the time for you to begin using the concepts introduced in the text to interpret media reports of current affairs. Look through current newspaper reports and find one that addresses the politics of a particular place. What components of the definitions of place provided by Agnew and Massey can you find in the article? What contests are in play? How do collective identities such as race, gender, age, class, and nationality interact with the concepts identified by Agnew and Massey?

The politics of scale

The actions of individuals and groups of individuals range in their geographic scope or reach. It is this scope or reach that is known as geographic scale. Place is one geographic scale, defined as the setting of our everyday lives. But place is just one scale in a hierarchy that stretches from the individual to the global (Taylor and Flint, 2000, pp. 40–6). (Perhaps even these boundaries are too narrow; genetic material and outer space could arguably be seen as the geographical limits upon human behavior.) As a simple example, let's talk about economics. Well, do you mean one's own personal financial situation, the "family fortune" or lack of it, the local economy, national economic growth or recession, the economic health of the European Union (EU) or the NAFTA (North American Free Trade Agreement) region, or the global economy? Each of these scales represents a different level of economic activity, or transactions that define local economic health or the trade and investment that spans the globe.

Now that we have introduced scales as a form of hierarchy we need to show that they should be thought of not as separate or discrete but as connected (Herb and Kaplan, 1999; Herod and Wright, 2002). To illustrate the point, if all businesses were thriving, then all local economies would be booming, every national economy growing, and the global economy healthy. But of course this is never the case; the viability of a business is partially defined by the opportunities within its scope. The family-owned hardware store or photocopying franchise is dependent upon enough wealthy customers nearby. A global company, such as Honda or Nike, negotiates the differential opportunities for sales in different countries. In turn, the relative prosperity of individuals is related to the economic health of the businesses they work in and those businesses' national and global markets.

Political acts also negotiate scale. Protest can be enacted at the individual scale, by breaking laws seen by the individual as unjust or wearing clothes or tattoos that make a political statement. An action such as not singing a national anthem when it is demanded or expected is another example of political action at the individual scale. But protest can also involve vigils outside of, say, abortion clinics or protests at animal hunts or laboratories conducting tests on animals. These "localized" acts require individual commitment and are also often motivated by national campaigns aimed at influencing the national legislative process. Increasingly, protest politics does not stop at the national scale; abortion politics, for example, are a component of discussions over the form of US foreign aid as well as a component of the missionary activity of many churches.

The examples show that geographic scales, like places, are socially constructed or made by human activity. We wear certain clothes and act in certain ways to create our own persona. Political parties and social movements are formed and maintained by individual activity, whether it be the highly public and visible speeches of the leader or the "bake sales" and envelope-stuffing activities of committed members. As the scope of the geographic scale increases, it is harder to envision how they are socially constructed; but the everyday practices of paying taxes, maintaining national armed forces, politicking for the "national interest," and cheering on national teams in the Olympic Games or World Cup ensure the functional expression of a country, and the sense of national identity. In the workplace, we act to produce and consume products that are the outcome of economic activities from across the globe; unconsciously we reproduce the tea-leaf pickers in Sri Lanka, as well as the brokers who trade the picked leaves, the bankers who finance the plantations, and the advertisers who suggest the merits of having a "cuppa" on a regular basis. Though scales are made by human activity, the larger their scope the less aware we are of the implications of our actions, and their importance in sustaining operations at that scale.

Participating in elections is another example of how scales are constructed by people. By choosing to vote or not to vote, an individual chooses to become involved in a particular way with the political system, either validating it or not. The aggregate of individual votes in a particular constituency creates a political jurisdiction as a particular political locality: either a "safe" or a "contested" seat. In addition, the outcome of individual votes creates a national political system; either maintaining established democratic practices or forcing a change in the political system. For example, in 2011 regional elections in Hamburg were interpreted as having meaning for the sitting national Chancellor Angela Merkel, as her party lost heavily, which in turn had implications for the role Germany would play in emphasizing particular policies for the EU. Finally, the example of elections shows that scales, just like places, are contested. Individuals may well compromise their own beliefs by voting for a party; for example voting for a British party because of their views on membership in the EU despite being uncomfortable about, say, social or educational policies. Furthermore, the constituency scale and national scale are the very product of competing political parties.

Activity

Consider how your actions occur within a hierarchy of scales. If you are a member of a political organization (a party or a pressure group) or a church, a member of the military services, or the employee of a business, think about how you are connected to a small group, which is part of a bigger organization. Think about the influences upon you and the smaller group that stem from national and global events. Also, think about how your actions work up this hierarchy of scales to construct the organization, as well as the national and global scales.

The contested nature of scales requires us to think more closely about how scales are made by political actions. In so doing, we need to move further away from the idea of a clear and distinct hierarchy of scales. Though the idea of a hierarchy is useful in introducing scales, it quickly breaks down when looking at actual politics. Instead we see that many scales are implicated in any one event or action. For example, an act of terrorism, such as a suicide bombing, is something that can only be understood by combining the psychology and motivations of the perpetrator, and the local, national, and global contexts. Targeting hotels in Kabul, Afghanistan that host Western governmental and aid organization workers for suicide bombing attack is not simply an attempt to kill the individuals in the building but a component of the violent conflict to determine political control of the country and an act against foreign presence in the country related to the military and civil politics of US-led global anti-terrorism. Many scales are implicated simultaneously to understand the causes and implications of what is often mistakenly reported as a "local" event.

Regions and politics

The region has been a staple ingredient for geography since it became an academic discipline. The idea that areas of the world can be thought of as homogeneous in such a way that they are different from other parts of the world is intuitive and useful. In terms of the physical world we readily make sense of the world by thinking of the desert region of Africa or the mountainous region of South America, for example. We can also think of regions at a number of scales. Within a country we often talk of the urbanized and rural regions. Within the Eurasian continental landmass the "post-Communist region" is often spoken of as an area that faces or produces different political, economic, and security issues than the region of "Western Europe." When speaking of inequality at the global scale, the regions of "Global North" and "Global South" have become the favored way to talk of regions that were once known as "the developed world" and "the developing world." As you can probably tell through all this use of quotation marks, defining regions is not as simple as it first appears. In fact, defining parts of the world as being within one region, and hence not in another, has always been a significant part of geopolitics.

Mapping regions is a component of "imagining" or representing the world in a way that creates differences. Such classification began in earnest with European exploration and the idea of a "New World" just waiting to be "discovered." Along with that came a categorization of the globe into "civilized" and "settled" regions (i.e. Europe) and other parts of the world that were "barbaric" or "open" and ripe for colonial control. In the twentieth century, the Cold War conflict was framed within a regionalization of the globe into First (the West), Second (Communist), and Third (the remainder) Worlds. The War on Terror has provoked regional categories such as the Axis of Evil, a particularly incoherent region, as well as "rogue states" and "safe havens" that imply regions at the scale of the state or parts of states that can be classified by lack of government control and, hence, insecurity.

Regions should be seen not only as a form of labeling or classification but also as the result of the construction of political institutions. This process creates what are known as functional regions. For example, the EU is a functional region defined by the spatial extent of membership and the reach of EU law and regulation. The construction of free trade zones, such as NAFTA or the Association of Southeast Asian Nations (ASEAN) in Asia, offers other examples of functional regions. The institutional creation of functional regions is not separate from the classification of formal regions. The EU is a very good example, as its functional extent has expanded eastwards to include countries such as Poland and the Baltic states, while Turkey has not been granted membership as some commentators debate its "Europeanness" – a polite way of saying that some do not want a Muslim country to be a member. As we shall see in Chapter 5, the result is a geopolitics of boundary definition and control that is a combination of delineating EU citizenship and defining a variety of outsiders as security threats.

Territory as political space

Just like region, the concept of territory is applicable and relevant at a variety of scales. Most definitions identify territory as a bounded space that is under some sort of political control. The most obvious example in geopolitics is a country, or more formally a state. Hence, territory is related to the political geography of sovereignty, or the idea of absolute power or control over an expanse of territory – just as we may declare a room in a house as "our space" that is off limits to others, as opposed to communal spaces such as the kitchen or living room. We also define territory through property rights. A piece of real estate is a demarcated territory over which the owner has control and authority and where others may be guilty of trespass. These examples illustrate that territory requires two related ideas that are essential to understanding geopolitics. First, territory requires some sort of political control. Second, politics, or the exercise of power, often requires territory. The necessary connection between these two points has been called territoriality – that power is exercised through the construction and management of territory (Sack, 1986).

For example, the ability of a government to exert power over its citizens requires the territorial demarcation of the country and an understanding that those within it are subject to the authority of a geographically defined government. Thus sovereignty, or the right to rule, is necessarily territorially defined: sovereignty is the right to rule within a specific and demarcated territory.

However, the territorial extent of sovereignty is never as clear as the definitions would like. On the one hand, we can think of areas of countries where government rule is weak or non-existent. For example, the ability for pirates to base themselves in Somalia is due to the non-existence of government control in the coastal region of Puntland. On the other hand, governments are often only partially sovereign within their own territory. The member states of the EU must behave in accordance with European laws enacted at a supra-state scale. Signatories to international conventions on torture or nuclear proliferation are, at least to some extent, restricted in their actions. We have also witnessed recently how the governments of Greece and Ireland have been forced to change their economic policy because of the poor evaluation of their debt commitment by financial institutions. Territory is both a fundamental building-block of geopolitics and something that is fluid over time and varies across space. We will explore the geopolitics of territory more fully in Chapter 4 when we investigate nations and states.

Activity

Think of a number of "territories" that you live in. In what ways do different spaces within your home have different rules of access or behavior for different members of your family or household? Ask yourself the same question for your home town. Finally, think of how the world is regionalized in a way that explains or justifies different forms of behavior in different regions. Perhaps you can ask who is "included" and who is "excluded" to help you think about these questions.

The politics of networks

The final geographic concept we will introduce differs fundamentally from the others. The study of networks has risen dramatically recently, not just in geography but across all the social sciences. While space, place, region, and (to some extent) scale can be seen as territorial, networks are seen as means to transcend territory. Networks are collections of nodes that are linked together. The nodes could be many things: terrorist cells, political activists, countries, or businesses, for example. Linkages can also take many different forms (such as flows of migrants between cities, the movement of weapons from one terrorist cell to another, or the connection of security alliances between countries). Linkages can also be of different strengths, such as the varying volume of trade between two countries or the degree of commitment countries make to defend each other. On the one hand, networks are seen to transcend the geographies of places and spaces because they connect geographically separated nodes, and often through linkages that do not travel through or cannot be controlled by the government of countries. For example, movements of currency initiated by traders in offices in different parts of the globe are beyond the regulation of governments. On the other hand, all nodes are situated in some place or other. If we think of a network of states then the nodes are geographic entities. If we are thinking about terrorist networks the cells are physically located, even if temporarily, in some geographic location. Currency traders operate in business offices that are, usually, situated in a major city such as London, New York, or Beijing. Hence, networks are interesting because they simultaneously transcend territorial geographic entities while being made up of cells that are within places and spaces. We discuss the geopolitics of networks fully in Chapter 6 in an examination of terrorism and transnational social movements.

Activity

What networks are you part of? (Perhaps you don't need to put down your laptop or cell phone to consider this question!) How is your hometown situated within networks? Are some of these networks more visible than others, and are some old and no longer relevant while new ones are being established?

The concepts described so far are all components of human geography that can be used to look at many different topics. We will use them to explore geopolitics, and so it is probably about time we defined what we mean by that word.

What is geopolitics?

Geopolitics is a word that conjures up images. In one sense, the word provokes ideas of war, empire, and diplomacy: geopolitics is the practice of states controlling and competing for territory. There is another sense by which I mean geopolitics creates images: geopolitics, in theory, language, and practice, classifies swathes of territory and masses of people. For instance, the Cold War was a conflict over the control of territory that was provoked and justified through geographically based images of "the Iron Curtain" and the "free world" and the "threat" of Communism from the perspective of Western governments and the "imperialism" of America from the Soviet Union's view.

So how should we define geopolitics, in the contemporary world and with the intent of offering a critical analysis? Our goals of understanding, analyzing, and being able to critique world politics require us to operate with more than one definition.

First, we must note the connection between geopolitics and statesmanship: the "practices and representations of territorial strategies" (Gilmartin and Kofman, 2004, p. 113). For now, we will take a limited perspective on this definition – and note how states or countries have competed for the control of territory and/or the resources within

Box 1.2 Winston Churchill's "Iron Curtain" speech

Here are some excerpts from Winston Churchill's (who had been Prime Minister of Britain during World War II) famous Sinews of Peace speech, made in 1946, in which he identified the "iron curtain" that would divide Europe throughout the Cold War. Coming just after the allied victory in the war, in which Britain, the United States, and the Soviet Union fought on the same side, this was a rhetorical watershed in the public's awareness of the Cold War, and the identification of the Soviets and Communism as a threat to peace. Read the excerpts and find phrases that refer to (i) the control of territory by particular countries, and (ii) the rhetoric or language used to either justify such control or identify it as a threat.

A shadow has fallen upon the scenes so lately lighted by the Allied victory. Nobody knows what Soviet Russia and its Communist international organization intends to do in the immediate future, or what are the limits, if any, to their expansive and proselytizing tendencies. I have a strong admiration and regard for the valiant Russian people and for my wartime comrade, Marshall Stalin. There is deep sympathy and goodwill in Britain – and I doubt not here also – towards the peoples of all the Russias and a resolve to persevere through many differences and rebuffs in establishing lasting friendships. We understand the Russian need to be secure on her western frontiers by the removal of all possibility of German aggression. We welcome Russia to her rightful place among the leading nations of the world. We welcome her flag upon the seas. Above all, we welcome, or should welcome, constant, frequent and growing contacts between the Russian people and our own people on both sides of the Atlantic. It is my duty however, for I am sure you would wish me to state the facts as I see them to you. It is my duty to place before you certain facts about the present position in Europe . . .

From Stettin in the Baltic to Trieste in the Adriatic an *iron curtain* has descended across the Continent. Behind that line lie all the capitals of the ancient states of Central and Eastern Europe. Warsaw, Berlin, Prague, Vienna, Budapest, Belgrade, Bucharest and Sofia, all these famous cities and the populations around them lie in what I must call the Soviet sphere, and all are subject in one form or another, not only to Soviet influence but to a very high and, in some cases, increasing measure of control from Moscow. Athens alone – Greece with its immortal glories – is free to decide its future at an election under British, American and French observation. The Russian-dominated Polish government has been encouraged to make enormous and wrongful inroads upon Germany, and mass expulsions of millions of Germans on a scale grievous and undreamed-of are now taking place. The Communist parties, which were very small in all these Eastern States of Europe, have been raised to pre-eminence and power far beyond their numbers and are seeking everywhere to obtain totalitarian control. Police governments are prevailing

(Continued)

> in nearly every case, and so far, except in Czechoslovakia, there is no true democracy . . .
>
> The safety of the world, ladies and gentleman, requires a new unity in Europe, from which no nation should be permanently outcast. It is from the quarrels of the strong parent races in Europe that the world wars we have witnessed, or which occurred in former times, have sprung. Twice in our own lifetime we have seen the United States, against their wishes and their traditions, against arguments, the force of which it is impossible not to comprehend, twice we have seen them drawn by irresistible forces, into these wars in time to secure the victory of the good cause, but only after frightful slaughter and devastation have occurred. Twice the United States has had to send several millions of its young men across the Atlantic to find the war; but now war can find any nation, wherever it may dwell between dusk and dawn. Surely we should work with conscious purpose for a grand pacification of Europe, within the structure of the United Nations and in accordance with our Charter. That, I feel, opens a course of policy of very great importance.

them. At the end of the nineteenth century, the European powers indulged in an unseemly struggle for colonial control over Africa, what is known as the "scramble for Africa." In a more contemporary sense, the geopolitics of the War on Terror produced alliances between states and the deployment of troops in Afghanistan, Iraq, and in bases across Central Asia. Inseparable from these "practices" of fighting in Iraq and Afghanistan, for example, is the role of representation: The fight against "evil," the spread of "democracy," etc.

Second, geopolitics is more than the competition over territory and the means of justifying such actions: Geopolitics is a way of "seeing" the world. From a feminist perspective, geopolitics is a masculine practice, hence my use of the term states*man*ship in the previous paragraph. In the much quoted words of Donna Haraway (1998), the practices and representations of geopolitics have relied upon "a view from nowhere." As we will see soon, geopolitical theoreticians have made claims that they can view or understand the whole globe. In other words, they operate under the belief that the whole world is a "transparent space" that is "seeable" and "knowable" from the vantage point of the white, male, and higher-class viewpoint of the theoretician (Staeheli and Kofman, 2004, p. 4 referring to Haraway (1998) and Rose (1997)). Geopolitical theoreticians classify the world into particular regions while also defining historical trends. The feminist critique rests on the idea that all knowledge is "situated" and, hence, "partial." The very fact that the classical geopoliticians were from privileged class, race, and gender backgrounds in Western countries meant that they had absorbed particular understandings of the world; they were unable to know the *whole* world. In stark contradiction, their policy prescriptions rested upon the assumption and arrogance of being able to see and know the whole world and the essence of its historical development.

Figure 1.3 The Iron Curtain.

A third understanding of geopolitics results from the identification of "situated knowledge": geopolitics is not just a matter of countries competing against countries; there are many "situations" or, in other words, the competition for territory is broader than state practices. Geopolitics has come to be understood as much more than war and the building of empire. It can also include racial conflicts within cities, the restrictions upon the movement of women in certain neighborhoods and at certain times because of patriarchal laws and/or fear of attack, and diplomacy over greenhouse gas emissions, as a few examples. In other words, geopolitics is not the preserve of states; individuals, protest movements, non-governmental organizations (NGOs) such as Greenpeace and Amnesty International, terrorists, and private companies are all engaged (and always have been) in the control of territory, and so have struggled to represent it in certain ways. Following Gilmartin and Kofman (2004), geopolitics is the multiple practices and multiple representations of a wide variety of territories.

Following on from the third definition, we identify our fourth and final meaning of the word. Geopolitics has come to include *critical geopolitics* (Ó Tuathail, 1996), which is the practice of identifying the power relationships within geopolitical statements. What assumptions underlie phrases such as "the spread of free markets" or the "diffusion of democracy," for example? What are the consequences of such practices and representations? Who gains what, and who suffers? In other words, the phrases commonly used to justify state practices are put under critical scrutiny to see how they try to restrict our view of the world and promote a limited number of policies. By promoting interpretations of world events that are counter to dominant government and media representations, critical geopolitics aims to encourage anti-geopolitics: practices by individuals, groups of citizens, indigenous peoples, etc. to resist the control and classifications imposed by states and other powerful institutions such as the World Bank.

Contemporary geopolitics identifies the sources, practices, and representations that allow for the control of territory and the extraction of resources. States still practice statesmanship; in that sense we are still offered "all seeing" interpretations of the world by political leaders and opinion makers. But their "situated knowledge" has been increasingly challenged by others in "situations" different from the clubs and meeting rooms of politicians and business leaders. As a result, geopolitical knowledge is seen as part of the struggle as marginalized people in different situations aim to resist the domination of the views of the powerful. Feminist geopolitics has invoked the need for a "populated" geopolitics, one that identifies the complexity of the world, and the particular situations of people across the world, as opposed to the simplistic models of classic geopolitics and their simple explanations (Gilmartin and Kofman, 2004, p. 115).

Geography and geopolitics

A theme we have been dealing with since the Prologue and the initial introduction of classic geopolitical theories is that geopolitical ideas are examples of "situated knowledge" that construct images of the world in order to advocate particular foreign policies. The "situation" of the knowledge is both social and geographical. All the classic theorists were white Eurocentric males with conservative outlooks and a degree of social privilege. The benefit of using geographic concepts to investigate geopolitics is that we can gain an understanding of the why and what of situated knowledge.

"Situation" can be analyzed through Agnew's geographic framework of location, locale, and sense of place. The geopolitical theories at the end of the nineteenth century were created in a location (in Agnew's sense of the word) of the relative economic strength of Britain, Germany and the US that drove the theorists' respective perceived foreign policy needs. The institutional settings of universities, government, and policy circles nurtured and disseminated the knowledge the theorists created. For example, both Mackinder's Eurocentrism and Kennan's derisive views of the Third World were generated through their socialization in particular family, social, educational, and professional settings that, in combination, made up a geographic locale. In sum, the classic geopoliticians carried a definite sense of place regarding their own country and other parts of the world which was instrumental in formulating their geopolitical outlooks.

The theorists' classification of the globe into particular regions also reflected Agnew's framework. The strategic importance of a country or region was evaluated in terms of its location, both resource potential and strategic role. Despotism, colonial administration, and "free institutions" were the types of locale attributed to countries to define policies. Finally, in order to justify the policies, a sense of place had to be disseminated to the public, both the "goodness" and morality of one's own country, but also the threat and depravity of other countries. In other words, the classic geopolitical theorists constructed geographical images of the world (or maps of locations and locales) within their own place-specific settings. Our job now is to provide a framework for seeing how geopolitical actions are "situated" within the dynamics of the world.

Geopolitical agents: making and doing geopolitics

Up to now, I have referred to the actions of individuals and "groups of individuals." It is time to tighten up the language and answer the question, who or what conducts geopolitics? In social science parlance, we will identify geopolitical agents. By agency, we simply mean the act of trying to achieve a particular goal. A student is an agent; her agency is aimed at completing her degree. A political party is an agent; its agency is aimed at seeking power. A separatist movement is an agent; its agency is targeted toward achieving political independence. A country may also be seen as an agent; its agency is seen in its trade negotiations, for example.

In the nineteenth and throughout most of the twentieth centuries, geopolitics was viewed as the preserve of the state (or country) and statesmen, with the gender referent being important (Parker, 1985; Agnew, 2002, pp. 51–84). Geopolitics was the study, some claimed science, of explaining and predicting the strategic behavior of states. States were the exclusive agents of geopolitics. But the contemporary understanding of geopolitics is much different; indeed, one set of definitions would classify all politics as geopolitics, in a broad understanding that no conflict is separate from its spatial setting.

Hence we can talk of corporations involved in the geopolitics of resource extraction as they negotiate with governments for mineral rights and maintain security areas within sovereign countries, or the geopolitics of non-governmental organizations (NGOs) seeking refugee rights, or the geopolitics of nationalism, as a separatist group uses electoral politics and/or terrorism to push for an independent nation-state, for example. A provisional list of geopolitical agents could include: individuals, households, protest groups, countries, corporations, NGOs, political parties, rebel groups, and organized labor, though this list is far from complete. Similar to our discussion of geographic scale, it follows that these agents are not separate but entwined: an individual is a member of a household, a citizen of a particular country, and may be affiliated with a number of political organizations, as well as being employed within a firm. Thus, not only does an individual act out a number of geopolitics, the geopolitics may be competing.

Geopolitical agents work toward their goals, but their chances of success and the form of their strategy is partially dependent upon their context. They do not have freedom of choice, but they do have choices. They also do not act within a geopolitical vacuum; they make calculations based upon other agents.

Let us look at two examples. First, Iran's decisions regarding the pursuit of nuclear weaponry are made in a calculation of the power of other countries, two of which are also nuclear powers, Israel and the United States. In this example, the geopolitical agent is identified as a nation-state or country (Iran), and its calculations involve awareness of other countries, or agents of the same geographic scale. Second, we can consider how South Korea's President Lee Myung-bak's decision to limit any military response to North Korea's cross-border shelling that killed four people in November 2010 was made after calculating the response of members of his political party, the parliament, and the South Korean electorate. In this example the actions of the leader of a nation-state (President Lee) required recognition of actions, or future actions, of agents at lower geographic scales, the political party and individuals.

Geopolitical agents can be thought of as geographic scales. Moreover, the way that geographic scales are connected to each other, and no event can be seen to be confined to one discrete scale, allows us to think of geopolitical agents as consisting of other agents and acting "below" or within yet more geopolitical agents. Our next conceptual task in this chapter is to explore what we mean by the use of the words "consisting of" and "within" in the previous sentence. We will do so through the terms structure and agency.

Structure and agency: possibilities, constraints, and geopolitical choices

The ideas of structure and agency are part of an intellectual debate within social science that can get us into some very complex philosophy. My goal here is to provide enough material for you to interpret contemporary geopolitics, rather than negotiating the philosophical debate. Provided below are some key rules to initially aid our discussion:

- Agents cannot act freely, but they are able to make choices.
- Agents act within structures.
- Structures limit, or constrain, the possible actions of the agent.
- Structures also facilitate agents, in other words they provide opportunities for agents to attain their goals.
- An agent can also be a structure and vice versa.

See Johnston and Sidaway (2004, pp. 219–64) and Peet (1998, pp. 112–93) for more on the theory of structures and agents and structuration theory.

What is a structure? A structure is a set of rules (formal as in legally enforceable laws) and norms (culturally accepted practices) that partially determine what can and cannot, could and should not, be done. In this sense, structures are expressions of power as they define what is permissible and expected. Agents are those entities attempting to act. In other words, a woman homemaker may be viewed as an agent, and the patriarchal household a structure. In another view, the very same household can be seen as an agent negotiating the laws and culture of a country, which is interpreted as the structure. And to take this further, that self-same country may be seen as an agent operating within the structure of the international state system with its international laws and diplomatic customs.

Why is this theoretical framework useful? First, it shows that agents face both opportunities to act but also constraints to their possible actions given the structures they operate within. For example, a labor union may have the legal ability to strike, but the same legislation may prevent blockading roads and other forms of civil disobedience. Second, agents will be able to use, and be frustrated by, a number of structures simultaneously, given the multiplicity of spheres they operate within. The labor union must also use friendly political parties and combat those that are critical, too. Third, we can see that a particular structure is not monolithic but made up of a number of agents. For example, the union consists of individuals who must take into consideration the needs of their own household. Hence, strikes can crumble as some union members vote for a return to work as financial pressures mount. No structure can be seen to be monolithic. Fourth, by knowing that agents are simultaneously structures and vice versa, we can think of the opportunities of agents and the barriers they face, within a hierarchy of geographic scale.

Thinking of the structures within which agents are operating as a hierarchy of scale allows us to identify the key spaces that are being fought over. In other words, we can define both the politics and the geography and, hence, the geopolitics in question. The agency of pro-democracy protesters in Egypt, for example, illustrates the importance of the national space of Egypt and the diffusion of pro-democracy movements across the Arab world – two related structures that gave the opportunity for protest. The inability of the ousted government of President Mubarak to constrain the agency of the insurgents was partially a function of the inadequacies of the national political structure.

Finally, it must be stressed that structures are the products of agents. A pro-democracy movement is made by the actions of its members, and the actions of the pro-democracy movement plays a role in making the national space what it is. However, in addition, the relationship is recursive. Or in other words, the national situation structures, to some extent, the actions of the pro-democracy movement while those actions construct the nature of national politics.

Activity

Reconsider how you located yourself within a hierarchy of scales in the previous exercise. In what way are you prevented from doing certain things because of norms and rules established at higher scales? In what way do norms, rules, and capabilities at higher scales allow you to do what you want to do? Also, in what way do your actions construct the norms, rules, and capabilities found at the higher scales?

Geopolitics, power, and geography

Geopolitics uses components of human geography to examine the use and implications of power. Contesting the nature of places and their relationship to the rest of the world is a power struggle between different interests and groups. The spatial organization of society, the establishment and extent (both geographic and jurisdictional) of state sovereignty, is a continuing geopolitical process. The political aspirations and projects of geopolitical agents are won and lost within a structure of geographic scales. The fortune of geopolitical agents is also a function of their component parts, which can also be seen as geographic scales.

Scale, place, and space are arenas, products, and goals of geopolitical activity, and each of those three concepts has many different manifestations. Already, we have seen that a wide variety of geopolitical agents can be identified. In sum, it can be seen that conflicts over places, spaces, and scales are pervasive and multi-faceted. To keep this book focused and manageable, particular forms of geopolitical conflicts and particular geographies will be emphasized. Though this is necessarily exclusive, I also encourage you to explore other forms of geopolitics.

Geopolitics, as the struggle over the control of spaces and places, focuses upon power, or the ability to achieve particular goals in the face of opposition or alternatives. In nineteenth- and early twentieth-century geopolitical practices, power was seen simply as the relative power of countries in foreign affairs. For example, in the early 1900s US naval strategist Alfred Thayer Mahan's categorization of power was based upon the size of a country, and the racial "character" of its population, as well as its economic and military capacity. In the late twentieth century, as the geopolitical study of power became increasingly academic, scholars created numerous indices of power, which remained focused on country-specific capabilities of industrial strength, size, and educational level of the population, as well as military might. Definitions of power were dominated by a focus on a country's ability to wage war with other countries. This traditional under-standing sees power as a material capability or resource, something that is possessed such as nuclear weapons, that allows one geopolitical actor to exercise power over another. In other words, power is forcing, or having the potential to force, another actor to act the way you want because of your material capabilities, such as military strength.

However, recent discussions of power have become more sophisticated and are critical of seeing power as a "thing." Instead, a relational sense of power is seen as more useful (Allen, 2003). Material capabilities only have an effect when two actors form a power relation. In other words, strong requires weak, or dominant requires controlled. Hence, social relations, and the abilities of actors to force, cajole, or convince another actor to do what is wanted, or for that "acted-upon" actor to resist, to varying degrees. For example, the power relations of nuclear proliferation lie not solely in the technical capacity to build a bomb but in the power relations inherent within some states being members of the United Nations Security Council and others, such as Iran and North Korea, being labeled rogue or outlaw states.

The growing emphasis on relational power was reinforced by feminist geopoliticians emphasizing that the focus on government capabilities ignores other forms of power, such as gender and racial relationships within and between countries that are, over time, assumed to be "normal" or of secondary importance to the male-dominated practices of foreign policy. Feminist insistence on the integral role of gender relations in geopolitics leads to connections between the competitive nature of power relations between countries and the way patriarchal relations within countries normalize a masculine and militarized conception of foreign policy (Enloe, 1983, 1990, 2004). Feminism forces us to think about the gender and racial make-up of geopolitical agents and structures, so promoting the

Figure 1.4 Woman and child in Iraqi bombsite.

study of geopolitics as the combination of multiple power relations. The result is that any understanding of a current event must come from a variety of perspectives and not just the calculations of male-dominated elites.

One of the other contributions of feminist and critical geopolitics analysis is the focus upon how power relations become taken for granted or viewed as "common sense." Power in this sense is the ability or need not to force others to do what you want, but to make them follow your agenda willingly without considering alternatives. These ideas, that we can call ideological power, stem from the Italian Marxist Antonio Gramsci (1971) who noted how a ruling class in a country needs to exert force to control the working classes only rarely. On the whole, subordinated groups "follow" political goals that are of greater benefit to the more powerful; alternatives are seen as "radical" or "unrealistic," while the dominant ideology is seen as "unpolitical" or "natural." For example, in the arena of international economics, policies for "economic development" created by the rich and powerful countries are adopted by the poorest countries of the world under the label of "progress" despite the growing global inequality levels after decades of such policies. The Gramscian notion of power requires us to consider how geopolitical practices and ideas are disseminated and portrayed to wide audiences in order to justify them and make them appear "normal" while belittling alternative views. In other words, the representation of geopolitics is another manifestation of power (Ó Tuathail, 1996).

Material, relational, and ideological power can be understood through considering the geographic concepts we have introduced. From Agnew's three aspects of place, location helps us understand material capabilities, institutions reflect and enforce dominant power relations, and sense of place would promote a "common sense" of what sorts of political behavior are the norm. Massey's definition of place would also encourage us to think how power relations construct places, especially power relations that use networks to transcend places and spaces. In another example, some geopolitical actors have the capacity to operate at different scales; for example, some politicians use local offices (such as being mayor or provincial governor) to develop political relations that allow them to become national political figures.

In this book, I will use the geographic concepts described in this chapter to understand how a variety of geopolitical actors use material, relational, and ideological power. The classic geopoliticians of the late nineteenth and twentieth centuries expressed confidence in knowing "how the world works" and used a historical-theoretical perspective to suggest or justify the foreign policy actions, mainly aggressive, of their own countries (Agnew, 2002). My goal is not to explain away the acts of any country or other type of geopolitical actor as the inevitable consequence of a deterministic world history. Instead, geopolitical agents and their actions are understood through examining the competition with other agents at a variety of scales, from local to global. Countries are an example of just one geopolitical agent, comprised of others, and interacting with other countries, non-state organizations, and multiple-state organizations (such as the North Atlantic Treaty Organization (NATO) and the United Nations (UN)) within a geopolitical structure. The complex interaction of agents and structures can be conceptualized as operating within fused or inter-locked geographic scales. All structures and agents are dynamic, their form and purpose contested. Such contestation requires us to think about different expressions of power, such as military capability and patriarchal relations, and their connections, in addition to the manner in which they are made to appear "normal."

Having read this chapter you will be able to:

- Understand the concepts of place and scale
- Understand the concepts of structure and agency
- Be able to think about places in the world as being unique and interconnected
- Be able to think of current events occurring within a set of interconnected scales
- Be able to think of current events as being performed by geopolitical agents
- Begin to consider how the actions of geopolitical agents happen within structures
- Consider the multiple forms of power that underlie geopolitics

Further reading

Agnew, J. (2003) *Geopolitics*, London: Routledge.

A more in-depth and theoretically sophisticated discussion of geopolitical practice and the way it has changed.

Cresswell, T. (1996) *In Place/Out of Place*, Minneapolis: University of Minnesota Press.

Develops and exemplifies the politics of place and identity, or the political geography of inclusion and exclusion.

O'Loughlin, J. (ed.) (1994) *Dictionary of Geopolitics*, Westport, CT: Greenwood Press.

An excellent resource for clarifying geopolitical terminology and also provides brief discussions of many geopolitical thinkers.

Ó Tuathail, G., Dalby, S., and Routledge, P. (1998) *The Geopolitics Reader*, London and New York: Routledge.

A collection of short essays providing easy access to many of the authors and documents introduced in this text.

Staeheli, L. A., Kofman, E., and Peake, L. J. (eds) (2004) *Mapping Women, Making Politics*, New York and London: Routledge.

An excellent collection of essays describing the feminist approach to the topics of geopolitics and political geography.

Flint, C. and Taylor, P. J. (2011) *Political Geography: World-Economy, Nation-State, and Locality*, sixth edition, Harlow: Prentice-Hall.

An introduction to world-systems analysis (discussed in the Prologue) as well as the broad content of contemporary political geography.

2

GEOPOLITICAL AGENCY: THE CONCEPT OF GEOPOLITICAL CODES

In this chapter we will:

- Introduce the concept of geopolitical codes
- Define the component parts of geopolitical codes
- Outline how geopolitical codes operate at different geographic scales
- Show that geopolitical actors other than countries also construct geopolitical codes by using the example of al-Qaeda

At the very outset of talk of war upon Iraq in 2003, there was little doubt that Great Britain would be the United States' most active and loyal ally. This was not a matter of force. The British government certainly had the choice to play a minor role in the conflict, or even try to use diplomacy to challenge President Bush's plan. Yet somehow it was "understood" that Great Britain would give political and military aid to its established ally. The British government's decision illustrates the features of the geopolitical actions of countries that we will discuss in this chapter: a country may choose to make particular foreign policy decisions, these choices are limited to varying degrees, and a partial influence on the choices made is the history of allegiances and conflicts.

In the previous chapter we introduced the concept of structure and agency. The goal of this chapter is to focus upon countries as geopolitical agents: the manner in which they make decisions within the global context. We continue the themes of geographic scale and structure and agency to interpret how countries make foreign policy decisions within regional and global contexts.

Geopolitical codes

The manner in which a country orientates itself toward the world is called a geopolitical code. Each country in the world defines its geopolitical code, consisting of five main calculations:

(a) Who are our current and potential allies?
(b) Who are our current and potential enemies?
(c) How can we maintain our allies and nurture potential allies?
(d) How can we counter our current enemies and emerging threats?
(e) How do we justify the four calculations above to our public, and to the global community? (Taylor and Flint, 2000, p. 62)

For example, Great Britain has defined its primary allies within the transatlantic and trans-European institutions of NATO and the EU. Furthermore, it has tried to retain influence across the globe through the establishment of the Commonwealth, made up of ex-British colonies. The latter has had mixed success: for example, the expulsion of Zimbabwe from the Commonwealth for its brutal campaign against white farmers in the face of strong criticism from Britain. The identification of enemies is also dynamic. Almost overnight, as the Soviet Union became Russia, it quickly changed from intractable enemy to ally. Another example is the actions of Brunei, Cambodia, Indonesia, Laos, Malaysia, Myanmar, Philippines, Singapore, Thailand, and Vietnam to establish ASEAN in a move to foster inter-state connections and counter-balance the regional power of China.

Attempts to maintain allies take a number of forms. Economic ties are one chief plank. The EU evolved out of relatively modest beginnings to integrate the economies of France and Germany and so cultivate a peaceful Europe after the brutality of the two World Wars. Cultural exchange is also another vehicle for maintaining or nurturing peace. Educational scholarships such as the Rhodes, Fulbright, and Goethe fellowships encourage international understanding and long-term ties. Business organizations such as the Rotary Club

Figure 2.1 US troops in Kosovo.

Box 2.1 Power and US Army relationships across the world

US Army Regulation 614 10 is entitled United States Army Personnel Exchange Program with Armies of Other Nations. It is a long, dry, bureaucratic document filled with awkward phraseology. It describes a policy and the rationale behind it. The objectives of the program are listed as:

(a) Establish on a mutually agreeable basis, relationships between US Army personnel and the personnel of armies of other nations by which experience, professional knowledge, and doctrine of the respective armies are shared to the maximum extent permissible within existing policies.
(b) Foster in the personnel exchanged and in their coworkers a mutual appreciation and understanding of the policies and doctrines of their respective armies through the sharing of professional knowledge and experience.
(c) Encourage the mutual confidence, understanding, and respect necessary to enable harmonious relationships to exist between the US Army/Government and the armies/governments of other nations.

"Harmonious relationships" is the desired outcome, not just at the scale of individual officers but between countries. There is an obvious military benefit in having officers of allies being able to work closely together, especially as the conflicts in Iraq and Afghanistan have been prosecuted by a coalition of armed forces. However, the program is more than that. It is a component of the US's geopolitical code, one of the means by which allies are made and maintained.

The full document can be found at www.apd.army.mil/pdffiles/r614_10.pdf. What form of power can you discern from the text? Think especially of the Gramscian and feminist definitions of power from Chapter 1. To answer this question, think about what norms and values the document promotes.

are also aimed at establishing linkages. The choice of "good-will" visits for incoming Presidents and Prime Ministers is indicative of which international relationships are deemed most worthy of attention (Henrikson, 2005). For example, it is a tradition that the incoming US President meets with his Mexican counterpart at an early date.

Military connections are also seen as a means to maintain international cooperation. NATO is perhaps the strongest case, in which it is determined that an attack upon one member is considered an attack upon all. Another means of connecting with allies is the sale of military equipment that is expected to tie the (normally) weaker buyer to the more powerful seller. However, there is no guarantee of subservience. Weapons supplied to Iraq during its war with Iran were subsequently seen as threats by the sellers, the United States and Great Britain. Less overt are the relationships fostered by military training (see Box 2.1).

Means to counter enemies are also varied. A once dominant but now, seemingly, outdated ingredient of the United States', the Soviet Union's, and Great Britain's geopolitical codes during the Cold War was appropriately named MAD, for mutually assured destruction. Nuclear capability was strong enough to annihilate enemies many times over. Of course, most of this weaponry remains. The belief was that, as destruction was assured, no one would dare start a nuclear war and "peace" would reign. At the other end of the spectrum is diplomacy: negotiations between governments to, at the least, prevent hostilities and, at best, nurture more friendly relations.

Sanctions are a common non-military means to force enemies to comply with one's wishes. An international campaign of sanctions and boycotts put pressure upon the South African government to end its apartheid policies. In 2010 the United Nations Security Council resolved, though not unanimously, to impose trade restrictions that targeted Iran's nuclear program, especially its ability to make nuclear fuel. Sanctions are often criticized for making the population suffer through lack of food or medical supplies rather than the politicians who formulate the policies in question. Countries can also change their opinion on the efficacy of sanctions: the British government under Margaret Thatcher disparaged the use of sanctions against apartheid South Africa; the governments of John Major and Tony Blair were strong advocates of sanctions against Iraq.

The fifth element of a country's geopolitical code should not be underestimated. The definition of an enemy, especially when it entails a call to arms, is something that can destabilize a government and lead to its fall. For example, intensifying the EU in the name of European peace and prosperity has proved equally exhausting for British governments. In another example, South Korean governments are constantly calculating how their actions and attitude toward North Korea will affect their popularity in future elections. In the wake of the November 2010 cross-border shelling by North Korea that killed four

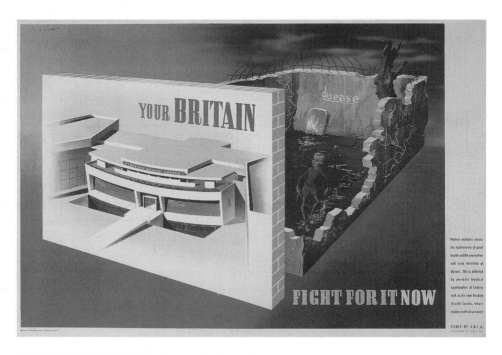

Figure 2.2 British World War II propaganda poster.

South Koreans, the opposition Liberal Democratic Party labeled President Lee Myung-bak as weak and ineffective.

Representational geopolitics is the essence of the fifth element of a geopolitical code. If enemies are to be fought, the basis of the animosity must be clear, and the necessity of the horrors of warfare must be justified. Enemies are portrayed as "barbaric" or "evil," their politics "irrational" in the sense that they do not see the value of one's own political position, and their stance "intractable," meaning that war is the only recourse. As we will see in the next chapter, these representations are tailored for the immediate situation, but are based upon stories deposited in national myths that are easily accessible to the general public.

Scales of geopolitical codes

Every country has a geopolitical code. For many countries their main, if not sole, concern is with their immediate neighbors: are they friends or enemies? Is increased trade or imminent invasion the issue? But some countries profess to develop a regional geopolitical code in which they have influence beyond their immediate neighbors. China's calculations toward expanding influence in Southeast Asia are a good example, as are Iran's attempts to further its influence in the Middle East and the Arab world (Taylor and Flint, 2000, pp. 91–102).

Finally, some countries purport to have global geopolitical codes. In the twentieth century the United States made geopolitical calculations based on a sense of national interest that required presence and action in all the regions of the world. In Chapter 7 we will talk more about states with global geopolitical codes by labeling them as "world leaders" (Modelski, 1987). At the moment it is sufficient to note that such a global geopolitical code is based upon a sense that their global presence is wanted. A challenge to their authority anywhere on the globe requires a response, for their legitimacy is based upon their global reach (Falah and Flint, 2004). On the other hand, world leadership requires world "follow-ship." Much diplomatic energy is spent to make sure countries are "on-board" the world leader's agenda. Any attempt by another country to create a global geopolitical code is interpreted as a challenge to the world leader. The growing influence of China within Africa and Iran's role in the Middle East are examples of how challenges within particular parts of the world are seen as challenges to the global calculations of the United States.

Though we can distinguish the power and influence of a country through designating its geopolitical code as local, regional, or global, it is false to separate local geopolitical codes from the global geopolitical context. Though the range of geopolitical calculations may be local, the influence of the global geopolitical context remains. For example, Hungary's decision to join NATO involved calculations about ethnic Hungarians in neighboring countries, and a future threat from Russia, but was still framed within the global authority and agenda of the United States (Oas, 2005). Hungary saw the changes in the global geopolitical context, as the world leader exercised its authority in Europe with the collapse of the Soviet Union, as an opportunity to advance its own security. The same idea can be applied to the way the "stans," the republics of Central Asia, utilized the War on Terror to obtain military aid from the US.

Box 2.2 Constructing threats: the CIA's view of the future

In November 2008, the National Intelligence Council, a group of "analysts" who synthesize "expert" advice and report to the head of the Central Intelligence Agency (CIA), released a report entitled "Global Trends 2025: A Transformed World." This report is not a geopolitical code in itself, as it provides scenarios based on "intelligence" rather than policy itself. However, it provides a basis of authority, the type of power that is viewed critically from Gramscian and feminist perspectives, which will likely underlie the revision of the US's existing code.

The report is notable for its emphasis upon economic and demographic change, especially the economic growth of China and India and demographic changes within Middle Eastern countries, and combined the implications for trade and oil consumption. As a result, perhaps we may see a return to the construction of Asia as an economic threat to US power that was predominant in the 1990s. Furthermore, terrorism inspired by Islamic fundamentalism is identified as a continuing threat, but so are the consequences of global climate change. The report foresees the US remaining as the dominant superpower, but its influence is partially diminished as European integration solidifies, the voice of other countries in international institutions increases, and Brazil, Russia, India, and China increase their stature and role. Instability in the Middle East, Sub-Saharan Africa, and Latin America is also raised as a concern.

Perhaps as interesting as the predictions the report contained was the manner in which they were presented. Particularly striking was the use of fictional scenarios within the report to give it urgency and accessibility. One scenario was a fictional letter from the SCO – Shanghai Cooperation Organization (an alliance, in the scenario but not in the current world, between Russia, India, and China, seeking dialogue as an equal partner with NATO). In the letter the formation of the SCO is explained as the result of the military withdrawal of the United States and its Western allies from Afghanistan and Central Asia as a result of military failure and declining defense budgets. Another scenario is a fictionalized 2024 article from the *Financial Times* declaring that the governments of states "are no longer king" and have been eclipsed by transnational networks. Climate change also makes a prominent appearance with a 2020 diary entry by the President of the United States lamenting how severe flooding resulting from a hurricane disrupted a meeting of the United Nations General Assembly in New York, and noting that its impact on re-establishing the New York Stock Exchange will be greater than the terrorist attacks of September 11, 2001.

Though these scenarios are within the realms of possibility, their sophisticated artistic representation blurs the boundaries between fiction, entertainment, and objective analysis. The scenarios may drive policy – whatever happens in Central Asia, for example. And though the SCO does exist it does not include India amongst its members. In this document, the means of representation and the

(Continued)

identification of the threat are blurred into a fictional presentation of potential threats and enemies.

The report is accessible at www.dni.gov/nic/PDF_2025/2025_Global_Trends_ Final_Report.pdf. Reading the report allows for the consideration of many of the concepts we have discussed so far. In what way does the report construct particular regions of the world? Does it focus upon particular scales to the relative exclusion of others? In what way does this report exemplify a focus on particular power relations and agents that feminists would criticize (especially the report's box on "Women as Agents of Geopolitical Change")? From a Gramscian perspective, what is "taken for granted" in this report and what are the implications? Is the representation of the report's findings successful in justifying its content?

Activity

Many countries post key foreign policy documents on the web. Choose a country of interest to you and see to what extent their code is aimed at the local, regional, and global scales and how they are connected. Also, identify how both force and diplomacy are combined in the statements.

A do-it-yourself case study: decoding the geopolitics of Central Asia

During the Cold war period and prior to its break-up the Caspian Sea basin and Central Asia were firmly under the geopolitical control of the Soviet Union. However, the past couple of decades have witnessed related changes that now make this region an arena of geopolitical contestation (O'Lear, 2004). Oil and gas reserves in the Caspian Sea basin are viewed as essential components of world trade, especially as economic growth in China and India drive up global demand. Yet, the size and accessibility of the reserves are disputed. Moreover, there is much geopolitical dispute over pipeline routes to transport the products through the politically volatile Trans-Caucasus region consisting of Georgia, Armenia, and Azerbaijan; the battle being over whether a route favoring European or Russian control is built (a geopolitical concern brought to the general public's attention via the James Bond movie *The World Is Not Enough* (Dodds, 2003)).

In addition, the recent establishment of the "stans" (Kazakhstan, Kyrgyzstan, Tajikistan, Turkmenistan, and Uzbekistan) has added further political tensions. Kazakhstan and Turkmenistan have their own oil and gas reserves, but these countries are also seen as pivotal in the War on Terror as they are portrayed as a "battleground" between Islamic fundamentalist groups and the establishment of "free institutions," meaning nominal democratic practices (though corruption and lack of political openness are rife) and free markets.

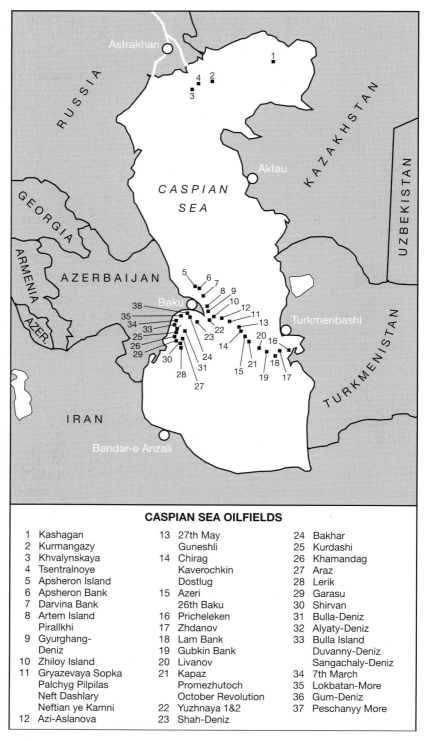

CASPIAN SEA OILFIELDS

1	Kashagan	13	27th May	24	Bakhar
2	Kurmangazy		Guneshli	25	Kurdashi
3	Khvalynskaya	14	Chirag	26	Khamandag
4	Tsentralnoye		Kaverochkin	27	Araz
5	Apsheron Island		Dostlug	28	Lerik
6	Apsheron Bank	15	Azeri	29	Garasu
7	Darvina Bank		26th Baku	30	Shirvan
8	Artem Island	16	Pricheleken	31	Bulla-Deniz
	Pirallkhi	17	Zhdanov	32	Alyaty-Deniz
9	Gyurghang-	18	Lam Bank	33	Bulla Island
	Deniz	19	Gubkin Bank		Duvanny-Deniz
10	Zhiloy Island	20	Livanov		Sangachaly-Deniz
11	Gryazevaya Sopka	21	Kapaz	34	7th March
	Palchyg Pilpilas		Promezhutoch	35	Lokbatan-More
	Neft Dashlary		October Revolution	36	Gum-Deniz
	Neftian ye Kamni	22	Yuzhnaya 1&2	37	Peschanyy More
12	Azi-Aslanova	23	Shah-Deniz		

Figure 2.3 The Caspian Sea.

Sources: US CIA; O'Lear (2004).

The final ingredient in the mix is the role of three powerful countries, each seeking to play a role in the region: China, Russia, and the United States. China and Russia claim "traditional" or "established" influence in the region. In other words, they resort to previous geopolitical codes to justify contemporary ones. For the United States, its increased presence in the region is couched in terms of the War on Terror, with access to oil and gas underlying the concern.

Political dynamics within countries after gaining independence, religious and political agendas, oil and gas, and the presence of three countries with a history of antagonism combine to make a complex situation.

Below are extracts from media reports on the geopolitical codes of China, Russia, and the US as they pertain to the republics of Central Asia. Read the extracts with an eye to the five elements of geopolitical codes.

Extract 1:

August 16, 2004, President George W. Bush introduced a radical plan to redeploy 60,000–70,000 US troops from European and Asian bases staffed for the Cold War to countries in Central Asia. A "senior defense official" was quoted as saying, "In the case of Uzbekistan, we have cooperation with them today on the war on terrorism. And we have believed that the war on terrorism will be with us for a period of time. And the kind of cooperation that develops further with Uzbekistan and others in Central Asia really depends on those countries to the extent they want to work with us."

There were both positive and negative reactions to this announcement. According to some it would inflame anti-American sentiment in the region; for others the potential redeployment was linked positively to the possibility of economic and political reform.

"US Redeployment Plan Sends Ripples through Central Asia,"
CDI Russia Weekly, August 24, 2004. www.cdi.org/russia/
320-16.cfm. Accessed January 14, 2005

Extract 2:

Central Asia and the Caucasus are emerging as the new focal point of rivalry between Russia and the United States in the wake of the Iraqi crisis. At the heart of the new standoff are rich oil and gas resources in the Caspian Sea basin, which may hold 100 billion barrels of oil alone. Washington has already a firm foothold in the local hydrocarbon industry, with US and joint US–British companies controlling 27 per cent of the Caspian's oil reserves and 40 per cent of its gas reserves.

In the post-Iraq scenario, the US has moved to put a bigger foot in the South Caucasus and Central Asia. It has mounted a titanic effort to revive GUUAM, a moribund economic and security group of five former Soviet states, Georgia, Ukraine, Uzbekistan, Azerbaijan, and Moldova, set up as a counterweight to Russian influence.

In Central Asia, the Pentagon has recently assumed talks with Tajikistan on the lease of three military bases in addition to the two the US established in

Uzbekistan and Kyrgyzstan in the wake of the 9/11 attacks. Last year [2002], Tajikistan received $109 million in economic aid from the US and accepted its offer to renovate a runway at Dushanbe airport.

> "A new Big Game in Central Asia," by Vladimir Radyuhin *CDI Russia Weekly*, from *The Hindu* posted July 18, 2003. www.cdi.org/russia/268-12.cfm. Accessed January 14, 2005

Extract 3:

Quoting a report by the Institute of Foreign Policy Analysis, "The United States should not allow itself to sort of get baited into a Cold War or a Great Game perspective on its relationship with China and Russia," Sweeney says. "Those states have legitimate interests in seeing Central Asia stabilized and in defeating Islamic extremism. So as long as their actions don't conflict with our core objectives in the war on terrorism, we don't need to be overly suspicious or reactionary to Russian or Chinese moves in Central Asia."

Last October [2003], Russia opened an air base at Kant in Kyrgyzstan to provide an air component for a rapid deployment force that will operate under the aegis of the Collective Security Treaty Organization (CSTO). The CSTO is a partnership among Russia, Armenia, Belarus, Kazakhstan, Kyrgyzstan, and Tajikistan.

Last year, members of the Shanghai Cooperation Organization – which groups China, Russia, and four Central Asian republics – held joint military exercises in Kazakhstan and China.

In 2002, China and Kyrgyzstan conducted a joint military exercise on the border areas of the two countries.

> "Central Asia: Report Calls on US to Rethink its Regional Approach," by Antoine Blua RFE/RL, *CDI Russia Weekly*, February 13, 2004. www.cdi.org/russia/13feb04-13.cfm. Accessed January 14, 2005

Extract 4:

Visiting the Tajik capital Dushanbe last month [October 2004], Russian president Vladimir Putin surprised his audience by pledging substantial financial investment in the Central Asian republic.

"The Russian side – both its state structures and private companies – intends to invest some $2 billion in the Tajik economy within the next five years . . . " Putin said.

And Tajikistan was not the only Central Asian country to receive promises from Moscow in October. Uzbekistan and, to a lesser extent, Kyrgyzstan, have been promised investment by Russia."

> "Central Asia: Russia Comes on Strong (Part 1)," RFE/RL, *CDI Russia Weekly*, November 19, 2004. www.cdi.org/russia/331-17.cfm. Accessed January 14, 2005

> ### Activity
>
> Activity: from the extracts, pick out the web of alliances that are being created in Central Asia and the potential tensions. As a starting point, one may want to look at the actions of Kyrgyzstan and the potential tensions between China, Russia, and the US. What means of generating allies are being used in this situation? In what way is the context enabling and constraining the actions of the countries involved?

The global geopolitical codes of the US

Tracing the story of the United States from the latter half of the nineteenth century shows how a country went from having a local code to a regional and then global code. As it recovered from a bloody civil war, becoming impressively urbanized and industrialized in some parts, while remaining "undeveloped" in others, the notion of expansion became a key issue in American geopolitics at the end of the 1800s. Despite much political debate, control of the Caribbean, and the Pacific became the focus of the United States' geopolitical code. Rear-Admiral Alfred Thayer Mahan was the theoretical light behind the US's move to globalism. Especially, he noted that sea-power was the basis for world power, but he was also careful to caution that any expansion of US influence would have to be done in a way that did not interfere with Great Britain's agenda and provoke war.

US national ideology was, and still is, based upon the rhetoric of anti-colonialism and national self-determination from British rule. Hence, especially at the beginning of the process of expanding the geographic scope of its influence, there was much domestic accusation that the country was embarking upon a policy of European-style imperialism unsuitable for the United States. But expansion did follow, and key geopolitical achievements were the defeat of Spain, control of Cuba, Hawaii, and the Philippines, and the construction of the Panama Canal. Related was reinforcement of the Monroe Doctrine that defined the US's sphere of influence across Central and South America, but also delineated, in an attempt to avoid conflict, that Great Britain and the United States each had distinct and exclusive realms of control across the globe (Smith, 2003).

Such geographic limitations were inadequate after World War II and the global role that the United States defined for itself in the face of the ideological and territorial challenge of the Soviet Union. The US created an unabashedly global geopolitical code. Table 2.1 illustrates how in 1947 the United States was including countries across the globe in its geopolitical calculations. In addition, policy toward particular countries was a function of "national security" and the US's "mission" to counter Communism.

NSC-68, written under the administration of President Harry S. Truman in 1950, is the key document outlining the new global geopolitical code of United States (NSC, stands for the National Security Council established by President Truman to serve as a forum to advise the President on foreign policy). It is useful in demonstrating the geographic imperatives of a global geopolitical code, as well as showing the similarities with the foreign policy of President George W. Bush, when the global presence and self-proclaimed mission of the US was facing a challenge from organized terrorism.

Table 2.1 Constructing a global geopolitical code

Identifying the global mission		Adding the national interest	Ranking the world
Threat from Communism		US security	Prioritizing countries
9	Great Britain	1	1
6	France	2	2
U	Germany	3	3
3	Italy	7	4
1	Greece	10	5
2	Turkey	9	6
7	Austria	6	7
U	Japan	13	8
10	Belgium	4	9
12	Netherlands	5	10
17	Latin America	11	11
U	Spain	12	12
5	Korea	15	13
U	China	14	14
13	Philippines	16	15
18	Canada	8	16
4	Iran	U	U
8	Hungary	U	U
11	Luxembourg	U	U
14	Portugal	U	U
15	Czechoslovakia	U	U
16	Poland	U	U

Source: The data is from a Joint Chiefs of Staff document reproduced in Etzold and Gaddis (1978, p. 79 and pp. 82–3), and the table is slightly modified from Taylor (1990, p. 16).

Note: U = unranked.

NSC-68 outlines the goals of a global geopolitical code, but it had to do so in the face of the geopolitical challenge of the Soviet Union and its ideological alternative, Communism. The document is oft-quoted for its claim that "The assault on free institutions is world-wide now, and in the context of the present polarization of power a defeat of free institutions anywhere is a defeat everywhere" (Section IV, A). The geopolitical implication of this statement is that all parts of the globe held equal strategic importance – the US believed it had to assert its authority in all countries. The Soviet system was a value system "wholly irreconcilable with ours" (IV, A), and its influence was preventing the establishment of "order" in the international system. Foreshadowing Chapter 7, NSC-68 claims that the conflict with the Soviet Union "imposes on us, in our own interests, the responsibility of world leadership" (IV, B).

The justification of the global geopolitical role of the United States was made clear: "Our overall policy at the present time may be described as one designed to foster a world environment in which the American system can survive and flourish. It therefore rejects

---

Activity

Activity: Compare the language of NSC-68 with the contemporary document Global Trends 2025 we discussed earlier in the chapter. How do the enemies and allies identified vary? Do the means of engaging allies and enemies differ or remain the same? In what way do they differ in how they refer to the global role of the US?

---

the concept of isolation and affirms the necessity of our positive participation in the world community" (VI, A). In other words, the simultaneous needs of defending a national "American system" but also diffusing it across the world were the basis for the US's global geopolitical code. The enemy was identified as the Soviet Union. Allies were countries and people advocating "free institutions."

The means of the geopolitical code were twofold. First, NSC-68 claimed a "policy to develop a healthy international community" (VI, A), a global geopolitical agenda in other words. Second, the document outlined a "policy of 'containing' the Soviet system" (VI, A), or negating the ideological and geopolitical challenger. Containment was a policy "which seeks by all means short of war to (1) block further expansion of Soviet power, (2) expose the falsities of Soviet pretensions, (3) induce a retraction of the Kremlin's control and influence, and (4) in general, so foster the seeds of destruction within the Soviet system that the Kremlin is brought at least to the point of modifying its behavior to conform to generally accepted international standards" (VI, A). The US would be the influential investigator, judge, and jury when it came to breaches of "international standards," but this policy manifested itself in realms of activity from nuclear deterrence, to the Vietnam War, and espionage. There is a contradiction within NSC-68. On the one hand, it calls for the global role and presence of the United States, while, on the other hand, its call for "containment" acknowledges the challenge of the Soviet Union. In other words, the rhetoric of leading the whole world was maintained within the practical constraints of a bipolar world.

But what of representing this global geopolitical code to domestic and international audiences? For domestic consumption, NSC-68 was based upon the ideals and content of the US Constitution. Section II was entitled "Fundamental Purpose of the United States" in which the "three realities" of individual freedom, democracy, and determination to fight to defend the American way of life were established and deemed to be under the protection of "Divine Providence." It was these "realities" that formed the basis of US world leadership; they were to be diffused to the world to maintain order. Section III, "Fundamental Design of the Kremlin" ("Design" having an evil, even sexual, implication rather than the valiant "Purpose") argued that the United States was the Soviet Union's "principal enemy." Both the domestic security and the global mission of the US were justified by rhetoric within NSC-68. The justification was also evident in the popular media. Hollywood produced a spate of movies based on biblical epics that portrayed the Middle East in a manner that was accessible while subtly justifying US foreign policy in the region (McAlister, 2001). Also, and as we will discuss in the next chapter, *Reader's Digest* published a series of articles to explain why the United States "had to" play a global role.

Geopolitical codes are dynamic. To illustrate this point we skip to the US geopolitical code in the wake of the terrorist attacks of September 11, 2001. The code remains global but the enemies and means are very different.

The War on Terror as a geopolitical code

How did the US respond to the attack on its global agenda and presence that was clearly manifested in September 2011? As in NSC-68, the United States focused on two separate but related geopolitical agendas: protection of its sovereign territory and the construction of a global order. The defining document was the National Security Strategy of 2002, the foundation of what became known as the "Bush Doctrine."

The National Security Strategy (NSS) is an annual exercise that updates the United States' geopolitical code. After the terrorist attacks of 9/11, the understandable focus was upon anti-American terrorism. By making the claim that the "struggle against global terrorism is different from any other war in history" (NSS, 5), the document was able to make the case that the established means to counter allies was ripe for change. The geopolitical threat identified by the NSS contained an apparent vagueness, but was able to become fixed on particular countries quite easily. The Strategy formalized the geopolitical code of the War on Terror, a war against "terrorists of a global reach" (NSS, 5). Simultaneously, this threat justified the global role of the United States while also laying the foundation for action against specific countries: The "enemy is not a single political regime or person or religion or ideology. The enemy is terrorism" (NSS, 15). The clever use of "not a single" allows the code to be nebulously global and also, at times, geographically specific.

The vague and the specific were combined in the identification of the threat posed by "rogue states", countries that "brutalize their own people and squander national resources" (NSS, 9). Such acts are deemed a violation of the "basic principles" and goals of a US global agenda that had been set in NSC-68. But rogue states are identified as a more specific threat too, being linked with the sponsorship of terrorism and the procurement of weapons of mass destruction. In this way, the notion of "rogue states" is able to give specific geographic definition, or targeting, to the global practices of the US (Klare, 1996).

With terrorism defined as the geopolitical threat facing the US, the "preemptive attack" was introduced as the legitimate means of countering the threat. The NSS evoked the United State's "right to self-defense by acting preemptively against such terrorists" (NSS, 6); simply to strike before "our enemies strike first" (NSS, 15).

In language that echoed NSC-68, the War on Terror was global in scope and historic in its intentions: "a global enterprise of uncertain duration" (NSS, opening statement). In another similarity with NSC-68, allies were to be maintained through "lasting institutions" (NSS, opening statement), that would provide the basis for "a truly global consensus about basic principles [that] is slowly taking shape" (NSS, 26). The intention was to secure the continuation of the global role of the US; "these are the practices that will sustain the supremacy of our common principles and keep open the path of progress" (NSS, 28).

However, such institution and agenda building was not deemed sufficient. The NSS includes means other than institutions and "principles" to secure allies. Indeed, now "is

Box 2.3 The geopolitics of the "Washington Consensus"

The geopolitical agenda and power of the US is as economic as it is military. Its influence in the key global economic institutions of the World Bank, International Monetary Fund (IMF), and World Trade Organization (WTO) is a reflection of its material interests and power to disseminate an ideological agenda. Indeed, since the 1990s the term "Washington Consensus" has developed as a summary of the economic policies that the US has pushed other countries to adopt, with much success. Under the umbrella of the term are policies of trade and investment liberalization, privatization, deregulation, fiscal and tax policy, and changes in the direction of public spending. Over time, those critical of such policies have also added issues of corporate governance, corruption, labor policy, and social safety nets into the argument.

"Washington Consensus" Global Trade Negotiations Home Page. www.cid.harvard.edu/cidtrade/isues/washington. html. April 2003. Accessed January 17, 2005

In combination, these policies, whether they are seen positively or negatively, fall under the phrase "Washington Consensus"; the economic side of the US's global geopolitical code.

In what way can the Washington Consensus be seen as separate or directly connected to the military actions of the United States?

the time to reaffirm the essential role of American military strength" (NSS, 29). But notably, the geography of this military strength was a global mission rather than the securing of the United States' borders: "The presence of American forces is one of the most profound symbols of the US commitment to allies and friends" (NSS, 29). Similar to NSC-68, the language of NSS balanced an identification of a threat to the US society and people, in terms of continued terrorist attacks, with a global commitment to promoting a particular vision of order. On the one hand, such order was deemed to be globally beneficial, yet, on the other hand, it was "a distinctly American internationalism that reflects the union of our values and our national interests" (NSS, 1).

The justification for the geopolitical code invoked language that was similar to that used in NSC-68: personal freedom was the goal, free-market economics was the means. The justification targeted toward domestic and global audiences: "A strong world economy enhances our national security by advancing prosperity and freedom in the rest of the world" (NSS, 17). The Strategy promoted free trade as the economic vehicle, a policy that was portrayed as having benefits for everyone across the globe: "This is real freedom, the freedom for a person – or a nation – to make a living" (NSS, 18).

In a related statement, made at a time of confidence after the "victory" in Afghanistan that led to the removal of the Taliban regime by an American invasion as punishment for their support of al-Qaeda bases, President George W. Bush used his annual State of the

Figure 2.4 "Freedom Walk."

Union speech to make focused geopolitical goals, within the framework of the War on Terror's global order. An "Axis of Evil," comprising Iran, Iraq, and North Korea, was identified. The geopolitical threat posed by these states was not just their alleged ties to terrorism, but also the identification of programs to build nuclear, chemical, and biological military capacity – weapons of mass destruction.

> Our second goal is to prevent regimes that sponsor terror from threatening our friends and allies with weapons of mass destruction. Some of these regimes have been pretty quiet since September the 11th. But we know their true nature. North Korea is a regime arming with missiles and weapons of mass destruction, while starving its citizens.
>
> Iran aggressively pursues these weapons and exports terror, while an unelected few repress the Iranian people's hope for freedom.
>
> Iraq continues to flaunt its hostility toward America and to support terror. The Iraqi regime has plotted to develop anthrax, and nerve gas, and nuclear weapons for over a decade. This is a regime that has already used poison gas to murder thousands of its own citizens – leaving the bodies of mothers huddled over their dead children. This is a regime that agreed to international inspections – then kicked out the inspectors. This is a regime that has something to hide from the civilized world.
>
> States like these, and their terrorist allies, constitute an Axis of Evil, arming to threaten the peace of the world. By seeking weapons of mass destruction, these regimes pose a grave and growing danger. They could provide these arms to terrorists, giving them the means to match their hatred. They could attack our allies or attempt to blackmail the United States. In any of these cases, the price of indifference would be catastrophic.
>
> We will work closely with our coalition to deny terrorists and their state sponsors the materials, technology, and expertise to make and deliver weapons of mass destruction. We will develop and deploy effective missile defenses to protect America and our allies from sudden attack. And all nations should know: America will do what is necessary to ensure our nation's security.
>
> We'll be deliberate, yet time is not on our side. I will not wait on events, while dangers gather. I will not stand by, as peril draws closer and closer. The United States of America will not permit the world's most dangerous regimes to threaten us with the world's most destructive weapons.
>
> (State of the Union Speech, 2002)

This passage contains the elements of a geopolitical code. The threat is identified: terrorism, state sponsors of terrorism, and weapons of mass destruction. A coalition of allies will be built and maintained. The means of obtaining these goals are, given that this was a broad-ranging political speech, understandably vague. However, the notion of preemptive strikes and the identification of specific countries, especially Iraq, were made clear. The justification of these actions was most direct: famine in North Korea, Iraqi "mothers huddled over their dead children," and "unelected" Iranian fundamentalist leaders were the opposites of the global order of prosperity, freedom, and civilization that the world leader had established as its agenda. Note the phrase "But we know their true nature": a claim that the US has the ability to cast an all-knowing "God-like" eye across the globe. Public events, such as the "Freedom Walk," embed the geopolitical actions and the justifications for them in domestic places – making geopolitics a combination of "global" foreign policy and "local" everyday life (Figure 2.4).

Activity

To keep this chapter at a reasonable length I have had to limit the analysis of US foreign policy documents. The statements of Presidents Carter and Reagan may be especially useful for you to investigate, or even President Theodore Roosevelt at the beginning of the US's rise to power.

The contemporary codes of other increasingly important countries (such as Turkey, Russia, or China) would also be intriguing to explore, given that the domestic politics of each one is fluid at a time when the country is increasing the geographic scope of its influence.

Other examples of geopolitical codes

As mentioned at the beginning of this chapter, every country in the world has a geopolitical code. Of course, it would be impractical to discuss even a handful of these geopolitical codes. Rather, the hope is that you will use your understanding of the concept to explore the geopolitical codes of countries that interest you. Some countries of particular interest, though, are Russia, India, and China. Their codes will be introduced briefly.

Russia

Over the past ten years the geopolitical code of Russia has vacillated between cooperation and antagonism with regard to the West. In the wake of the terrorist attacks of September 2001, Russia cooperated with the United States in the War on Terror. One reason for this policy was a calculation that the new actions of the US would provide a free hand for Russia to enact against insurgents, labeled as terrorists, within the troubled province of Chechnya (see Chapter 4). However, rising energy prices emboldened Russia and it changed its stance to one of reasserting its authority in neighboring countries (which had either been part of the Soviet Union or within its sphere of influence during the Cold War). The key event was the brief war with Georgia in 2008. This war, and the antagonism it raised within the West, led to the recognition in Russia that economic cooperation with other countries was necessary for economic modernization, and without such development it would not have as much international influence as the US and EU. For a further discussion see Mankoff (2010).

India

After World War II, and gaining its national independence, India defined and practiced a significant geopolitical code that professed non-alignment with either of the Cold War superpowers (the US and the Soviet Union) and aimed to be a catalyst for Third World solidarity. The end of the Cold War initiated drastic changes, though. India changed its economy from one that was largely centrally planned to a free-market model open to investment and trade. Also, India established itself as a nuclear power in light of its long-standing antagonism with Pakistan and an eye to past hostilities with China and that country's own nuclear arsenal. With nuclear weapons and staggering economic growth, India no longer felt that it needed strength through a bloc of Third World countries but could act as a regional power in its own right. This ambition has been tempered by the ongoing territorial dispute in Kashmir, the continued rivalry with Pakistan, and two terrorist threats (one from extreme Muslim groups connected to Pakistan and the other from the internal Maoist Naxalist groups).

China

It is almost inevitable that changes in China's geopolitical code will dominate the next few decades. As with India, China's unprecedented economic growth has enabled a change in its geopolitical code. Its economic growth has made it a global investor, buying much of the US's national debt and gaining influence in Africa through financing development projects. It has developed its military to be modern and extending its reach, through nuclear power and the development of a deep-sea navy. Though it still claims Taiwan to be a breakaway province of China, trade, investment, and business people readily flow between the island and the mainland. China and Russia have formalized a new security relationship (with Central Asian states), encoded in the Shanghai Cooperation Organization. Its key role in global trade, and the economy of the United States in particular, has meant that it is simultaneously treated as a partner and friend while others, with a more hawkish eye, see it as a potential military competitor.

These three short paragraphs are just a glimpse into three particular codes. I encourage you to explore them in more depth. One thing to bear in mind is that all geopolitical codes can only be understood relationally: in other words, codes are made in relation to the codes of other countries, whether they are deemed allies or enemies. A collection of alliances between countries can create security regions, such as NATO, the EU, and the SCO. Also, networks of trade and diplomatic regions can also be used to secure peaceful relations. Although the actions of states, mediated by the geographies of regions and networks, may create peaceful relations, their geopolitical codes must increasingly recognize and incorporate non-state agents.

Geopolitical codes of non-state agents

So far we have presumed that geopolitical codes are relevant for one type of geopolitical agent, states. But when we introduced the idea of geopolitical agency we saw that states are just one of many forms of geopolitical agents. Do non-state geopolitical agents have geopolitical codes? Or is the concept of geopolitical code useful in understanding the actions of non-state agents? In this section we analyze the actions and statements of al-Qaeda, a very significant non-state agent even after the killing of Osama bin Laden, to explore these two questions before considering other types of non-state agents.

A geopolitical code to challenge the United States

In February 1998, the London-based Arabic language newspaper *al-Quds al-Arabi* published a statement signed by Osama bin Laden, and four other men prominent in radical Islamic politics. The statement opened with two quotes from the Koran before setting the geopolitical scene for its readers:

> The Arabian peninsula has never – since God made it flat, created its desert, and encircled it with seas – been stormed by ant forces like the Crusader armies spreading in it like locusts, eating its riches and wiping out its plantations. All this is happening at a time in which nations are attacking Muslims like people fighting over a plate of food. In light of the grave situation and the lack of support, we and you are obliged to discuss current events, and we should all agree on how to settle the matter.
>
> (Quoted in Ranstorp, 1998, p. 328)

The "current events" were portrayed as "three facts that are known to everyone" (Ranstorp, 1998, p. 328). The overarching theme was the "self defense of Muslims against aggressive forces," manifest in the US military presence in Saudi Arabia since the first Gulf War in 1991, and the cooperation of the Saudi regime with the United States (1998, p. 328). The *fatwa* was a call to arms, and such a "fact" was portrayed in vitriolic language: "[F]or over seven years the United States has been occupying the lands of Islam in the holiest places, the Arabian peninsula, plundering its riches, dictating to its rulers, humiliating its people, terrorizing its neighbors, and turning its bases in the peninsula into a spearhead through which to fight the neighboring Muslim peoples" (1998, p. 328).

This first "fact" was supported by two others: "the Americans are once again trying to repeat the horrific massacres" (1998, p. 329), in other words conflict with Iraq and other Muslim countries was about to intensify; and "the Americans' aims behind these wars are religious and economic, [but] the aim is also to serve the Jews' petty state and divert attention from its occupation of Jerusalem and murder of Muslims there" (1998, p. 329). Part of bin Laden's success has rested on the second "fact," continuing through the 2003 war on Iraq, giving the *fatwa* a prescient nature. The third "fact" cites an established geopolitical issue, the Israel–Arab conflict and the plight of the Palestinians: however, it introduces new geopolitical actors into this situation. The subtext is a claim that the leaders of Arab states have been unable to stop the Zionist policies of Israel, and so it is now up to al-Qaeda to make a stand rather than politics as usual.

Following the components of a geopolitical code, the enemy identified by bin Laden's *fatwa* is clear; it is the United States and its "crusade" against Muslims, plus its regional ally, Israel. Though the Saudi regime is strongly criticized, bin Laden "focuses on presenting the Saudi regime as a 'victim' whose dependent military and security relationship with the 'crusader forces' has led King Fahd to act subserviently to US interests and designs" (1998, p. 326). Rather than blaming particular groups or leaders, bin Laden was trying to identify a broad range of allies by emphasizing "unity." But these allies are nebulous, and certainly not a list of countries. Rather, the "we and you" that are called to arms suggests that it is interconnected individuals and groups who will enact the geopolitics of the *fatwa*.

The means of the geopolitical code were as simple as they were brutal. "We – with God's help – call on every Muslim who believes in God and wishes to be rewarded to comply with God's order to kill the Americans and plunder their money wherever and whenever they find it" (1998, p. 329). The means of maintaining allies rested in the interpretation of anti-US violence being the divine will of God; unity would come through "every Muslim" following God's will. The justification of such geopolitics was, for bin Laden, found within the Koran. The reference to divine will is the ultimate justification for action.

Bin Laden's *fatwa* was, of course, translated into horrific acts of terrorism that provoked the US invasion of Afghanistan, and the overthrow of the Taliban regime. More controversially, links between al-Qaeda and Saddam Hussein, partially retracted by President George W. Bush's administration, were used in the justification for the 2003 war on Iraq. For bin Laden, the US's presence in the Middle East was seen as evidence of "their eagerness to destroy Iraq, the strongest neighboring Arab state, and their endeavor to fragment all the states of the region such as Iraq, Saudi Arabia, Egypt, and Sudan into paper statelets and through their disunion and weakness to guarantee Israel's survival and the continuation of the brutal Crusade occupation of the peninsula" (1998, p. 329). The allegation was that the United States was acting as an imperial power, dividing in order to conquer. Statements made by al-Qaeda after the killing of bin Laden in May 2011 continue to stress the same geopolitical basis for their activity; focusing upon the presence of the US in the Middle East, its support for Israel, and its perceived hostility toward Islam. However, the reaction in Muslim countries to bin Laden's death was rather muted and did not provoke a series of mass rallies, suggesting that the movement toward peaceful democratization is a greater force than extremism and terrorism.

Other non-state agents and geopolitical codes

The terrorist attacks of September 11, 2001 brought the relevance of non-state actors into discussions of foreign policy: and not before time. Contemporary foreign policy documents are replete with talk of insurgents, failed states, rebel groups, organized criminal organizations, pirates, and their interactions with terrorist groups. These non-state actors are frequently portrayed as interacting and posing threats to states that must, therefore, be addressed in states' geopolitical codes. For example, in early 2011, senior officers in Britain's Royal Navy were complaining that proposed cuts to their budget would eliminate traditional patrols in the Caribbean aimed at the illegal narcotics trade.

Though the example of al-Qaeda easily identifies this particular terrorist group as displaying all the components of a geopolitical code, can the same be said for other non-state actors? Insurgents would seem to be a case where a geopolitical code is a useful tool to identify whom they are fighting and cooperating with, through what means, and for what reasons. However, are pirates and organized crime syndicates just out to make money and hence not in need of the strategy and representations that are the content of a geopolitical code? It would be easy to dismiss such groups as purely criminal, with no political agenda. However, pirates and criminals are often accused of cooperating and supporting insurgents and terrorists, and the membership of such groups may be shared, and hence the distinction becomes blurry. If what are ostensibly criminal groups see benefit in cooperating with politically motivated non-state actors, then perhaps they do calculate allies, enemies, and means to engage them.

Also, what about non-state actors that are not necessarily violent? Is it useful to think of actions of trans-national social movements, such as anti-globalization or environmental groups as being guided by their own geopolitical codes? They certainly identify targets (such as Japanese whaling fleets or meetings of international bankers); these movements are often coalitions of a number of groups (or allies); they develop increasingly sophisticated means of demonstration and disruption; and they are effective in telling the media of their motivations. If regions such as the contemporary Middle East, and the former Soviet Union and its sphere of influence in the 1990s, can be so dramatically altered by pro-democracy social movements, then it seems that their geopolitical impact requires a consideration of their geopolitical calculations, or codes.

Activity

Am I correct in arguing that the concept of geopolitical codes is applicable to a non-state agent such as al-Qaeda or a social movement? To justify your answer, think about how the means of maintaining allies and engaging threats must differ for geopolitical agents other than countries. Must representation differ too?

Summary and segue

Understanding the concept of geopolitical codes allows for an analysis of the multiple agendas that countries face and the diversity of policy options that are available to address them. Moreover, geopolitical codes are contested within countries as different political interests within a country seek different policies. Geopolitical agents do not have complete freedom in defining their code; the context of what other, perhaps more powerful, countries are doing must be taken into account. The dynamism of geopolitical codes is a result of the interaction, perhaps inseparability, of domestic politics and the changing global context. The idea of a geopolitical code may also be relevant in explaining the actions of non-state geopolitical agents. In the next chapter, we will concentrate on the fifth element of geopolitical codes, the way they are represented to gain public support.

Having read this chapter you will be able to:

▧ Define geopolitical codes
▧ Interpret government foreign policy statements as the manifestation of geopolitical codes
▧ Consider the actions of geopolitical agents other than countries as the manifestations of their geopolitical codes

---

Further reading

Flint, C. and Falah, G.-W. (2004) "How the United States Justified its War on Terrorism: Prime Morality and the Construction of a 'Just War'," *Third World Quarterly* 25: 1379–99.

A discussion of how the United States, as world leader, has different needs, and uses different language, in justifying its geopolitical code compared to other countries.

Halliday, F. (1983) *The Making of the Second Cold War*, London: Verso.

An excellent discussion of the actions of the United States and Soviet Union in the Third World that provides background for the discussions of US geopolitical codes.

Klare, M. T. (1996) *Rogue States and Nuclear Outlaws: America's Search for a New Foreign Policy*, New York: Hill and Wang.

Provides background to current geopolitical pronouncements regarding "rogue states" and the "Axis of Evil."

Ranstorp, M. (1998) "Interpreting the Broader Context and Meaning of bin Laden's *fatwa*," *Studies in Conflict and Terrorism* 21: 321–30.

As the title indicates, this article provides background and context for bin Laden's initial call for a conflict with the United States.

3

JUSTIFYING GEOPOLITICAL AGENCY: REPRESENTING GEOPOLITICAL CODES

In this chapter we will:

▨ Introduce the cultural aspect of geopolitical codes

▨ Focus on the ways in which geopolitical codes are justified

▨ Identify the linkage between popular culture and foreign policy

▨ Introduce the concept of Orientalism

▨ Discuss how popular culture helped the public interpret the content of NSC-68

▨ Discuss how the administration of President George W. Bush represented its War on Terror

▨ Discuss how Saddam Hussein used a combination of Arab nationalism and Muslim belief to justify his 1990 invasion of Kuwait

▨ Map the changing geography of US foreign policy representations in the State of the Union speeches

---

The previous chapter concentrated upon understanding the practices of states and non-state agents by using the concept of geopolitical codes. An essential dimension of a geopolitical code is the way that a country's decisions and actions are justified. A convincing case for why a country is a "threat" or not, and what should be done about it, must always be made not only to a country's own citizens, but also to the international community. This chapter will explore how violent acts of geopolitics (the prosecution of wars) are portrayed as the defense of a country's material interests plus its values.

The examples in the chapter include Hollywood movies, *Reader's Digest*, the speeches of Saddam Hussein, and the changing content of the US Presidential State of the Union speeches. The overall conclusions are that geopolitics is pervasive (we participate in geopolitics by being part of popular culture), and that geopolitical representations are fluid and adaptable to changing contexts.

War! What is it good for . . . ?

On the surface, the "Soccer War" between El Salvador and Honduras provides an illustration of how petty national concerns and hatreds can explode into warfare. The value of "national pride" was marshaled to provoke and justify a war. However, just focusing on national differences, in this case, is a shallow and incomplete understanding, as we shall see. In 1969 El Salvador and Honduras played two games of football ("soccer") in the qualifying stages for the 1970 World Cup finals (see Kapuscinski, 1992, pp. 157–84, for a full narrative of this conflict). The first game, in Honduras, resulted in a one–nil victory for the home side. Back in El Salvador, eighteen-year-old Amelia Bolanios committed suicide in light of the national shame. Her funeral was a national event, the procession led by the President of El Salvador and his ministers. The return match in El Salvador was played in an extremely hostile atmosphere; El Salvador won three–nil. The Honduran team retreated to the airport under armed guard; their fans were left to their own devices and two were killed as they fled to the El Salvador–Honduras border. The border was closed in a matter of hours. The Honduran bombing of El Salvador and military invasion followed shortly afterward. The war lasted 100 hours; 6,000 people were killed and 12,000 wounded; the destruction of villages, homes, and fields displaced approximately 50,000 people.

But are nationalist passions sparked by football matches enough to initiate the horrors of war? Underlying the tension between El Salvador and Honduras, a tension that easily aroused national hatred as footballs landed in goal nets, was a struggle for land and human dignity that crossed an international border (Kapuscinski, 1992, pp. 157–84). The land of tiny El Salvador, with a very high population density, was owned by just fourteen families. In a desperate attempt to obtain land, about 300,000 El Salvadorans had emigrated, illegally, across the border and established villages. The Honduran peasants also wanted land reform, but, and backed by the US, the Honduran government avoided redistributing land owned by its own rich families and the dominant United Fruit Company. To avoid an internal political struggle, the Honduran government proposed to redistribute the land that the El Salvadorans had settled. The prospect of forced repatriation from Honduras not only unsettled the migrants, but also rattled the government of El Salvador, who faced the prospect of a peasant revolt.

Landlessness, monopoly, human dignity, fear of popular rebellion; these mutual "domestic" issues were intertwined across the porous Honduras–El Salvador border. The government's decision to go to war was made within a context of class inequality and the inequities of land ownership. National humiliation on the football field was merely the fuse that lit the political tinderbox. International war was deemed a more obvious solution than altering the domestic status quo.

A more recent consideration of the role of material "needs" and ideological hype in oiling the movement toward war was evident in *Fahrenheit 9/11* and its portrayal of the geopolitical doctrine of President George W. Bush. According to filmmaker Michael Moore, the war upon Iraq was a matter of capitalist greed: the maintenance of personal fortunes built upon the global need for oil. Unsurprisingly, the reasons President Bush gives for going to war are different. They do not rest upon the material needs, and financial gains, of extracting and selling oil. Rather they rest in the realm of ideals. According to this interpretation, the war was fought in the name of "freedom" and securing the privileges of the citizens of the United States in the face of terrorist threat. Also, the ideals of "liberating" the Iraqi people from tyrannical rule and bestowing them with democratic self-rule were cited. As in the Soccer War, contemporary discussions of war juxtapose two, equally political, interpretations of the causes of war: material gain and values.

Answering the question headlining this section is clearly beyond my capabilities. But let me try and provoke an initial approach to the question in a way that provides some insights into particular conflicts while also placing war within our broader discussion of geopolitics. Our discussion will focus on two different reasons for fighting wars, specifically the reasons governments use to justify their involvement in conflicts: material interests and values. These two reasons should not be seen as competing or mutually exclusive. Instead, they are presented as the two most common themes used to justify participation in warfare.

A prime philosopher of the material motivations for war was V. I. Lenin (1939). For Lenin, the leader of the Bolshevik Revolution in 1917 and first Premier of the Soviet Union, the upcoming wars were materialist in nature, an expression of the imperialism of the rich powers needing new markets and sources of raw materials to feed the banks and finance groups within their borders. For Lenin, the two World Wars were the bloody component of the continuous struggle for profits. The Soccer War and the war on Iraq could be interpreted in the same way.

Alternatively, sociologist Pitirim Sorokin (1937) argued that war is fought over competing values. The national humility fatally felt by Amelia Bolanios in El Salvador in 1969 was a sign of the power of values in warfare. The impassioned speeches of President George W. Bush and Prime Minister Tony Blair regarding Iraqi liberation frame the invasion of Iraq, and other episodes of the War on Terror, as a conflict over values: values that are deemed to justify loss of life amongst coalition forces, humanitarian workers, insurgents, and Iraqi citizens.

Rather than attempting to portray, and resolve, a simple debate between a "materialist" and a "values" perspective on war, the aim of this section is to initiate an exploration of the different geographies of representation that result from the material and value interpretations of war. Representations of war that are based upon material concerns or "interests" are territorially based, often reflecting concerns over control of territory or boundary location in order to access key resources. On the other hand, representations of war that resort to ideals are less bound to specific pieces of territory, and tend to speak to visions of what is best, or "common sense" for humanity.

Activity

For any foreign policy event of your choice (a war, the imposition of sanctions, the establishment of alliances, etc.) look at policy documents, speeches, or media commentaries that portray the policy, and evaluate the degree to which justification was made through material interests or values. Are the relationships between material justifications and territoriality and value-based justifications and extra-territoriality that I posit evident?

Box 3.1 *Dulce et Decorum Est*

Bent double, like old beggars under sacks,
Knock-kneed, coughing like hags, we cursed through sludge,
Till on the haunting flares we turned our backs
And towards our distant rest began to trudge.
Men marched asleep. Many had lost their boots
But limped on, blood-shod. All went lame; all blind;
Drunk with fatigue; deaf even to the hoots
Of tired, outstripped Five-Nines that dropped behind.

Gas! Gas! Quick, boys! – An ecstasy of fumbling,
Fitting the clumsy helmets just in time;
But someone still was yelling out and stumbling
And flound'ring like a man in fire or lime . . .
Dim, through the misty panes and thick green light,
As under a green sea, I saw him drowning.

In all my dreams, before my helpless sight,
He plunges at me, guttering, choking, drowning.

If in some smothering dreams you too could pace
Behind the wagon that we flung him in,
And watch the white eyes writhing in his face,
His hanging face, like a devil's sick of sin;
If you could hear, at every jolt, the blood
Come gargling from the froth-corrupted lungs,
Obscene as cancer, bitter as the cud
Of vile, incurable sores on innocent tongues, –
My friend, you would not tell with such high zest
To children ardent for some desperate glory,
The old Lie: *Dulce et decorum est*
Pro patria mori.

Wilfred Owen, 1917

Cultured war

Dulce et decorum est pro patria mori (It is sweet and proper [or fitting] to die for one's country). It is only of late that Hollywood has begun to portray the horror, pain, loneliness, and indignity of dying in war. Movies such as *Platoon* told a story of the Vietnam War. Steven Spielberg's *Saving Private Ryan* was technically brilliant in showing the terror, confusion, and slaughter of the Normandy landings of World War II, but its main purpose was an act of remembrance and national thanks for the World War II generation:

Figure 3.1 World War II memorial, Stavropol, Russia.

supported by the book and film *Band of Brothers*. None of these efforts comes close to the cynicism of Wilfred Owen's poem; for Owen juxtaposes the brutality of individual death with the romantic mythology of nationalism. As the soldier Owen describes is feeling life slip away as his lungs are being corroded by gas, is he really going to reflect on the "sweetness" of his duty to give his life for his country? In actuality, the common cry of the dying soldier, usually a young man, is for his mother (Fussell, 1990).

Yet, at the beginning of World War I, millions of people across the European continent and within Britain greeted the outbreak of war with unbridled joy (Eksteins, 1989; Tuchman, 1962). People lined up to join their respective military; it seemed like a great thing to be going off to war. Owen's cynicism came later, and was a product of experience at the Front, and a reaction to what he saw as the inhumanity of nationalism driving young men to their death.

World War I is widely seen as the epitome of the modern war (Eksteins, 1989), but it also ushered in the rise of fascism, especially Adolf Hitler and the Nazi party. We may all be familiar with the term Nazi and Nazism, but it is important to reflect on the meaning of the name. "Nazi" stems from the full title of Hitler's party National Socialist German Workers Party. The "national" and the "socialist," or the emotive and the material, combined powerfully in Hitler's ideology to give one of the clearest, and most reviled, expressions of nationalism in history. Nationalism is the belief in a common culture, or people, and its connection to a particular country. The term will be discussed in greater depth in the following chapter. Hitler's rhetorical strength lay in his ability to link material grievance with an ideologically based future within a portrayal of the German national past and future glory. Though Nazism is an extreme, the extreme simply serves to illuminate what is common in contemporary geopolitics. Representations of war and other forms of geopolitics are usually based within an understanding of an individual's membership in a national group, which has its particular values, traditions, and history.

The rhetoric of Japan prior to World War II paralleled the language being used in Nazi Germany. The country, through the political efforts of the military and the Emperor, defined a "national defense state" that emulated the Nazis' complete commitment of society to the prosecution of war. The goal was to construct Japanese imperial control in China and other parts of Asia, and maintain colonial control of Korea. The calculation was to gain material resources to fuel Japan's economic modernization and growth, for material reasons. However, it was portrayed as a fight to prevent the influence of Western values and morals in Asia, as a racial enterprise to unify Asia (under Japanese dominance), and as an expression of the Shinto state-religious will, or as a "holy war" (Bix, 2000).

Geopolitical actions are given meaning in order to justify their prosecution. Geopolitics is, then, a cultural as well as a political phenomenon, and usually a national one. Culture normalizes the continuous prosecution of geopolitics across the globe. More specific-ally, it paints "our boys" (and to a much lesser extent "our girls") as heroes fighting a valiant and necessary fight, while portraying the enemy, or "them," as evil and villainous (Fussell, 1990; Hedges, 2003). Increasingly, "they" are made invisible – deaths that we need not worry about as we prosecute war (Gregory, 2004).

Me, a geopolitician? Laughable!

First, let us explore what we know without knowing, or at least without thinking or questioning. *Do not look at Box 3.2 until you have read the following lines.* First, make a column of numbers from 1 to 12. Second, get ready to look at the list of countries in Box 3.2. Don't look yet. I want you to read the name of each country in turn and write the first word that comes into your head – no matter what it is. The key of this exercise is not to think too deeply, and not to worry about what you are writing. Tip: don't write the name of the country, just move through the list quickly. *Go!*

Once you have written the list you can consider the following questions:

1   What are the sources of the images or ideas behind the word you wrote? Think of movies, news reports, books, lectures, magazines, songs . . . which have created a picture of a country for you – one positive and one negative.

2   Do these images reflect particular groups in society? In other words, do you think the image comes from a male or female perspective, a white or other racial position, an elite or non-elite group?

3   What are the implications of these images to the foreign policy of your country? In other words, do these particular images and the response they generated in your mind facilitate particular policies?

4   Which terms or words that you came up with lead to justification of foreign policies that were either violent or required no action?

This simple exercise is trying to suggest that we all carry around "knowledge" of countries that we probably know very little about. This "knowledge" is gained from the most dubious sources, primarily Hollywood movies and television shows, and complemented by songs, jokes, and comedy routines, etc. It is nothing new. As a boy of about seven or eight years, I can remember my grandfather playing a recording to me by a comedy duo, Flanders and Swan. One of their songs was called something like "The English are best, I wouldn't give tuppence for all of the rest." It was a list of all the peculiar faults and traits that are possessed by different national groups, and in the process expunging any negative characteristics from each and every English person. This may seem harmless, but it is powerful because it is pervasive and everyday. Listening to the record of an evening was "family fun," that just happened to instill a belief that my country was obviously superior to any other. Such "humor" was the basis for a geopolitical understanding of Britain's "right" to tell other countries, using force if necessary, what to do.

My whole generation grew up in England on a steady diet of "Irish jokes": Continually painting an image of all Irish people as hopelessly stupid. How could I then, as I grew older, begin to think there was a historical basis for Irish nationalism? A deeper understanding of this conflict, and others, had to be actively sought by myself despite the obstacles of the "common knowledge" provided by mainstream media sources and cultural attitudes. My knowledge of the Irish had been created by the English media and the telling of Irish jokes at the back of the bus; what else did I need to know? The playground, the bus-stop, and the couch in front of the TV were very important arenas for an understanding of geopolitics. The basis for these images was not just schoolyard jokes passed down from older to younger siblings, but also the result of cultural products such as movies, books, magazines, and songs.

In the Gramscian sense of power, we carry with us "knowledge" of the world that is often of the must dubious and partial nature, but the knowledge is powerful nonetheless. Its power comes from its being taken for granted as "common sense," on the one hand, and in the way that knowledge is the foundation for the "ideals" used to justify geopolitical actions. For example, if whole swathes of the world are deemed "anarchic" then policies combining non-involvement in some cases (such as Rwanda) or military intervention in other cases (such as Afghanistan) may be implemented with little need to explain or defend them. Of course, production of the cultural common sense underlying foreign policy cannot be left to the imagination of playground humorists; the media industry is heavily implicated.

"Freedom," and *Reader's Digest*

In the previous chapter we discussed NSC-68 as a defining document of the US's geopolitical code. NSC-68 was not America's bed-time reading in 1950. The global geopolitical code of the United States had to be disseminated to the public in more appealing media. Throughout the twentieth century, *Reader's Digest* provided a source of geopolitical commentary for the American public. A widely read magazine with a broad and loyal audience, the *Digest* has captured the consistent and changing geopolitical messages that were aimed to help its readers understand, in order to support, the global geopolitical role the US had assumed (Sharp, 2000).

Box 3.2 Geopolitical word association

1   United States
2   North Korea
3   France
4   Mexico
5   Afghanistan
6   China
7   Turkey
8   Iraq
9   Japan
10   Somalia
11   Pakistan
12   Great Britain

A consistent theme throughout the century was the comparison of the US to classical societies who had given so much to humanity through the diffusion of civilization. In 1922 the *Digest* stated:

> The art that was Greece and the legal temperament that was Rome reflect the idealism of great peoples who had something within their national souls which became the common heritage of humanity. This is the supreme test of a truly great people ... No fear need be felt that the historians of the future will pronounce national humanitarianism – the will to disinterested human service – the original national contribution of the United States to the higher idealism of the world.
>
> (Sharp, 2000, p. 75)

In this quote, the global mission of the United States is portrayed as a benevolent act, disseminating a humanitarianism that will benefit all, similar to a mythic interpretation of the Romans, the provider of the rule of law, and not the efficient and technologically superior invaders of other countries. Note too that the "duty" and the ability of the US to conduct this global mission rest not in government and military power but in national characteristics. In other words, the individual American has a global role to play, not just the NSC, the Pentagon, and the President. *Reader's Digest* paid greatest attention to the Soviet Union in the 1950s and 1960s. In the 1980s and 1990s new threats were identified, including growth of "big" government and "political correctness" (Sharp, 2000, p. 339).

Two related threats to "America" were identified by *Reader's Digest*. On the one hand was the concern of degeneracy, or the fact that the US's power could wane, similar to Rome's classical fate (Sharp, 2000, p. 75). On the other hand was Communism, portrayed as a threat to humanity as well as to the culture and moral fabric of America, as the magazine argued in 1935 and again in 1950 and 1948:

Table 3.1 Reader's Digest *articles identifying threats to the American way of life*

	1986–8	1989–91	1992–4
Russia and Communism	48	50	34
Terrorism	6	8	1
Drugs	13	19	5
Japan/economic	7	9	10
Domestic danger	27	51	76
American values	13	11	25

Source: Slightly modified from Sharp (2000, p. 339).

Note: By "domestic danger" the *Reader's Digest* means the bureaucratic influence of big government, as well as the spread of "political correctness."

The American dream of Poor Boy makes Good leads even the most underpaid drudge to consider himself a potential millionaire. This makes it hard to arouse him to a Marxian consciousness.

(Sharp, 2000, p. 77)

The unsuspecting American imagines that we are safe from socialism because he knows the people will never vote for it. But socialism can be put over by a small minority.

(Sharp, 2000, p. 89)

The power of the Soviet Union, and particularly the Soviet Communist Party, is due to the fact that, while in a sense the Soviet state has moved into a power vacuum in Europe and Asia, the Soviet Communist Party has moved into a moral vacuum in the world.

(Sharp, 2000, p. 89)

Sharp shows how *Reader's Digest* was able to connect the US's global geopolitical agenda to the concerns and actions of the individual citizen. Wars in Korea and tensions in Berlin, for example, were represented as the necessary outcomes of a mission bestowed upon the US because of its national character. The Cold War was portrayed as a conflict over values: US humanitarianism versus Soviet totalitarianism. Moreover, the "battle" was not geographically distant. *Reader's Digest* made linkages across spaces and down geopolitical scales, so that an immediate threat to the fabric of society was constructed. The result was that geopolitical agency was essential for all Americans, whether they were fighting Communism in a foreign land or working as a farmer, teacher, shop assistant, etc. in Nebraska. The *Digest* made this clear in 1952:

This is where *you* come in. No-one is too small or insignificant, too young or too old, to be shackled and regimented, or pauperized and destroyed . . . By its all-encompassing timetable sooner or later [the "communist masterplot"] has to reach you.

(Sharp, 2000, p. 93)

Individual morality as a form of everyday geopolitical agency remained an important theme for *Reader's Digest* after the Cold War. Increasingly, "domestic danger" became the geopolitical focus of the magazine, as the threat of Russia and Communism declined through the 1990s (Sharp, 2000, p. 151). Continuing a theme that began with the New Deal program, in 1994 the *Digest* noted that "The barbarians are not at the gates. They are inside" (2000a, p.152): continuing the magazine's reference back to the classical ages of Rome and Greece, made accessible by the Hollywood biblical epics discussed earlier (McAlister, 2001). Who were the barbarians? Surprisingly, it was the American government itself, or more specifically the portrayal of an increasingly powerful central government and, in the argument of the *Digest*, a consequent culture of dependency upon government services. In themes that loyal readers would be able to trace back to early discussions of communism, government intervention (or degrees of communism) created the potential of moral decay.

In sum, *Reader's Digest* offered a commentary on US global geopolitics that assigned an exceptional nature to the country that demanded a global presence. Participation in wars across the globe, and other forms of political intervention and presence, were related to the everyday life experiences of the readers by a representation of the American nation as one embracing particular moral standards and norms. In this manner geopolitical agency was portrayed as a daily necessity for the individual; as communism and big government threatened American communities and, by extension, the US's ability to conduct its humanitarian mission for the whole world.

Jason Bourne: representing men of action and the national security state

The representation of the US in the world continued with the uncertainties of the post-Cold War period and the practices and rhetoric of the War on Terror. Increasingly the medium of television and film played the dominant role in popular culture justifications of the US. The work of geographer Klaus Dodds (2010) has highlighted the key representations in one series of films: those that portray the journey of CIA operative Jason Bourne. The sequence of movies gained popular and critical acclaim: *The Bourne Identity* (2002), *The Bourne Supremacy* (2004), *The Bourne Ultimatum* (2007). The trilogy tells the story of Jason Bourne who is discovered afloat in the Mediterranean Sea suffering from amnesia. It turns out he was part of an illegal CIA assassination squad. The films, in the midst of relentless breathtaking action sequences, tell his story of discovering what he did, and uncovering the plot despite the existence of senior CIA officers trying to cover up their wrong-doing, and kill Bourne in the process.

By looking closely at the Bourne films we can see some of the standard and pervasive representations that exist in Hollywood action movies and TV shows (Dodds, 2010), and are part of what has been called "the new violent cartography" in popular culture (Shapiro, 2007). A key message in the movies is the need for a pervasive and intrusive state in order to provide security. This message is sent by emphasizing two key themes: the portrayal of gender roles and the role of place. The national security state that is portrayed in the Bourne movies is dependent upon a largely male workforce that acts within a certain definition of masculinity. The male figures, epitomized by Bourne himself as portrayed by Matt Damon, are strong, protector figures who identify those who must be protected (usually women) and those who must be confronted because they pose a threat. Bourne plays a key masculine role because he acts in this highly masculinized way against the corrupt men of the CIA, epitomizing the national security state. "If Bourne's quest is deserving of viewers' sympathy, it is because of the fact that his male superiors are breaching their duty to secure national security for their domestic populations" (Dodds, 2010, p. 23). Furthermore, Bourne portrays other characteristics of the male protector: he is highly adept with an array of weaponry and demonstrates advanced fighting skills, and he is willing to take risks and immediate action against identified threats, often to the extent of killing people.

These masculine attributes, put to good use to protect the national security state, are enacted through some familiar elements of the action movie. The numerous chase scenes illustrate the ability of the national security state to track and pursue people in any place across the globe (Dodds, 2010). There is also the notion of a "race against time" in which the public and the state are aware of an imminent threat and so need, and want, the male heroic protector figure to carry out his plan of action (2010). These political and gender narratives take place against a global backdrop in which particular places have meanings to generate the audience's understanding of geopolitics. In the Bourne movies, Washington, DC and New York City are hubs for control and calculation that have global implications. Moscow and Berlin are contrasted as seedy legacies of the Cold War, and Tangiers is seen as an "Orientalized center of contemporary intrigue" (2010, p. 26).

What of the women in the movie? They do not only play the roles of those who must be protected. They also humanize Bourne by helping him figure out who he is and attempting to guide him back into a life outside the intrigue and violence of the national security state (though the woman whom he befriends in the first film and who becomes his girlfriend, is promptly shot at the beginning of the second for her efforts). One key character in the series is Pam Landy (a senior CIA figure) who becomes aware of the illegal covert operation of which Bourne was part. While her male colleagues are urging her to kill Bourne, to keep their illegal assassination team secret, Landy tries to understand Bourne's motivations while at the same time defining a national security state that can be both brutal and moral. Her male colleagues continually tell her that she is out of place in the men's world of national security, but she challenges them by showing she is a woman of action tempered by thought and understanding (Dodds, 2010).

Security, the state, gender, and geography all play a role in creating the geopolitical representations within the Bourne films, and similar representations appear in most contemporary Hollywood action films. They illustrate the casual way that we experience geopolitics and are the building blocks for the everyday understandings of the world that create our perceptions of the world and what is deemed necessary and appropriate geopolitical action. Yes, you cannot even go to the movie theater/cinema or turn on the television without being part of the construction of geopolitical codes!

In our discussion of the role of Tangiers in the Bourne movies we used the term "Orientalized." This is a key underlying concept in geopolitical representations and one we will now explore.

> ### Activity
>
> *Reader's Digest* and the Bourne movies are an example of both the Gramscian and the feminist definitions of power. Look at figure 3.2 and identify the gender roles that are portrayed. Discuss how the gender roles build upon our taken for granted assumptions about how foreign policy is conducted and by whom.

Our embassies abroad continue to hire local employees who are sometimes as adept at pampering our foreign-service people as they are at espionage

Why Our Embassies Are Nests for Spies

Condensed from
WASHINGTON MONTHLY

PRISCILLA WITT

AMERICANS WHO take assignments at U.S. embassies around the world have always been warned about foreign agents who—through seduction, bribery, bugging or other means—will attempt to recruit them to obtain classified information. In the wake of the recent Marine spy scandal in Moscow, the United States has witnessed just how real the dangers are.

Even more startling, we have seen that these spies needn't hang out in smoky clip joints, waiting to ensnare unsuspecting embassy employees. Instead, today's foreign agents can sometimes be found inside the protective walls of American embassies, as paid employees of the U.S. government.

Our Moscow embassy, which once had 220 Soviets on its payroll,

Figure 3.2 "Nests for spies."

Orientalism: the foundation of the geopolitical mindset

As we have seen, representations of geopolitics, with clear messages regarding personal and national behavior, are embedded within a whole host of media that "entertain" and "inform" us, without claiming to be overtly political. They are in movies and books, etc., but the presence of these representations in readily accessible media is the product of much deeper cultural structures that go under the title of Orientalism.

Edward Said (1979) was the driving force behind the concept of Orientalism, by which he meant the institutionalized portrayal of non-Western cultures as "uncivilized," "backward," "child-like," even "barbaric" and "primitive" in such a manner that it pervaded government, academic, and popular culture circles. Said was a professor of English Literature and analyzed novels, especially by English authors, of the nineteenth century. However, his work is still relevant today, and is the basis for many academic works on how "knowledge" of other cultures is created and disseminated. Furthermore, the point of Orientalism is that such "knowledge" of, say, Arabs, or Muslims, or Africans is a form of power. There is power in the ability of Western countries to create particular understandings of the rest of the world, or to classify weaker countries and their inhabitants. For example, Western media portrayals of African countries are pervasive; African representations of Europe and the US are not. Such knowledge becomes unquestioned because it is seen everywhere. Second, the authority of the knowledge, given that it is largely unquestioned or countered by alternative images, allows for, or demands, particular foreign policy stances toward particular countries. Orientalism is the foundation of the responses to the geopolitical word game we played earlier (Box 3.2). North Korea *is* nuclear weapons, for example.

Said, however, did not only point out that the West portrayed non-Westerners as barbarians to justify their colonization. There is a double-sided nature to the process too. By portraying non-Westerners as "backward" and "uncivilized," etc. Western countries and their geopolitical practices were painted, for self-consumption, as the exact opposite: "Modern," "the bearers of civilization," etc., and hence the "natural" rulers of the globe. This self-portrayal of the West was not done just to make people feel good about themselves: the extremely brutal acts of conquest and oppression that were necessary for the West to establish its imperial rule over the world could then be seen as the required, if unfortunate, acts needed to "discipline" or "civilize" the "natives." If the competitive colonization of Asia was known as the "Great Game" in a reference to the sports-field escapades of the British ruling class, then the household belief that "to spare the rod is to spoil the child" was also transferred to the global scale – in the belief that "natives" only understood discipline. Orientalism did not die with the end of formal Empire. In fact, it has been noted that the portrayal of vast numbers of human beings as "savages" and "barbarians" has been in resurgence in the wake of the terrorist attacks of September 11, 2001 (Gregory, 2004).

The *profession* of Orientalism, as Edward Said called it, continues today. Academics at respected universities write books and newspaper columns, and make television appearances, that combine to tell us the world out there is full of savage and irrational people just waiting to inflict pain and suffering on the innocent West. Kaplan's (1994) "The Coming Anarchy" is a good example, as is, with specific focus on Arab countries and Muslims, the work of Bernard Lewis (2002). Increasingly, the target of Orientalism has become Islam – a topic readily adopted by Western media. These everyday acts of portraying a dangerous world needing US policing, with the help of other Western countries and especially Britain, are established on the nearly 200 years' worth of cultural products first analyzed by Said. The contemporary catalyst was Samuel Huntington's (1993) "The Clash of Civilizations": epitomized by its classification of the world into eight "civilizations" – the most problematic one being Islam with its "bloody innards and bloodier boundaries." Empirical analysis does not support Huntington's bold claims – in statistical analyses of conflicts across the world, connections to Islam do not increase the likelihood of war (Chiozza, 2002). But who reads the academic journals? The talking heads and op-ed pieces are the "high-brow" contributors to "common sense" and "Indiana Jones" is the low-brow.

Activity

Consider the movie releases in your hometown over the past, say, six weeks. Who were the enemies or "baddies" portrayed in the movies? Do they represent, either overtly or subtly, real-world countries or other geopolitical agents? Who do the "goodies" represent? What were the nationalities of the actors who played the "goodies" and the "baddies"? Consider the gender roles in the movies; can you trace geopolitical messages akin to the interpretation of the biblical epics we discussed earlier?

Scholars were quick to point out the cultural misrepresentations in Huntington's work, but it still, along with the work of Robert Kaplan, sowed the seeds of a post-Cold War understanding that the world was "chaotic," "messy," and "dangerous" and hence needed "order" and "stability" (Dalby, 2003; Flint, 2001). Perhaps more insidious is the contemporary Orientalist practice of making whole populations invisible. With the invasions of Afghanistan and Iraq in the 2000s the language of Orientalism has changed significantly, according to Gregory (2004). Iraqi and Afghani people are dehumanized – either by making them invisible by just not mentioning them, or by portraying them as "savages," beyond our civilized codes and not deserving of political or economic support.

The new media representations of satellite images and computer simulations allow the Western viewer to be a virtual participant in the War on Terror. Geographer Derek Gregory (2004, pp. 197–214) talks of this development at length; noting the interactive websites of *USA Today* and the *Washington Post* that allowed you to point and click over Baghdad, retrieve "details" of the targets, and keep track of the war by seeing photos before and after the bombs were dropped. At the same time, the images were almost completely empty of pictures of human suffering and carnage. Someone in Birmingham, Alabama or Birmingham, UK could "repeat the military reduction of the city to a series of targets, and so become complicit in its destruction – and yet at the same time . . . refuse the intimacy of corporeal engagement" (2004, p. 205). Geopolitics has become just another computer game of killing the bad guys, only in this case the victims are not just computer-animated figures; they are absent. The essential point that Gregory is making by focusing on websites and computer games is that Orientalist representations are now something that the general public actively participate in and help create rather than being "fed."

At the same time, death and suffering is officially absent, in a breach from historic military practice – the deaths of enemy combatants and non-combatants were not counted in the invasion of Iraq and the subsequent insurgency. Gregory's use of blunt official statements is most effective: "We do not look at combat as a scorecard" and "We are not going to ask battlefield commanders to make specific reports on battlefield casualties" (2004, p. 207). From the Western perspective, contemporary war can be a computer game, just as long as you do not keep track of the human consequences: maybe it is the only computerized conflict available that does *not* allow you to count points!

However, in a time of electronic and globalized media, alternative visions are available. The Al-Jazeera satellite television company was broadcasting images of carnage in Iraq across the Arab world, showing pictures described by its editorial staff as "the horror of the bombing campaign, the blown-out brains, the blood-spattered pavements, the screaming infants and the corpses" (2004, p. 208). In a change from the nineteenth- and early twentieth-century Orientalist situation described by Said, the technology to broadcast the story of the victims is now possible. However, the legacy of Orientalism lies not just in the ability to broadcast, but in what gets seen and how it is interpreted. Here the Western powers still have some advantage. Images from sources other than main-stream Western media are easily dismissed as "cinematic agitprop," or stories reported "from the enemy side."

As technology and the global geopolitical context have changed, then so has the role of the public – becoming "embedded" in such a way that the creation and consumption of geopolitical information becomes blurred (Der Derian, 2001). News reports are now filled with, sometimes dependent upon, tweets and YouTube videos sent by members of the public at the scene of geopolitical events. For example, media coverage of the 2011 demonstrations in Arab countries was, at times, reliant on such forms of input. In a sense, the ability for the "colonized" to speak back and give their own version has been enhanced too. The Internet is a means for people across the globe to be given the perspective of what it feels like to be "liberated" (see Box 3.3). The representations of the most powerful countries' geopolitical codes, especially that of the US, are still touted and are powerful, but there are alternative interpretations too.

Box 3.3 Baghdad blog

For me, April 9 was a blur of faces distorted with fear, horror and tears. All over Baghdad you could hear shelling, explosions, clashes, fighter planes, the dreaded Apaches and the horrifying tanks tearing down the streets and highways. Whether you loved Saddam or hated him, Baghdad tore you to pieces. Baghdad was burning. Baghdad was exploding . . . Baghdad was falling . . . it was a nightmare beyond anyone's power to describe. Baghdad was up in smoke that day, explosions everywhere, American troops crawling all over the city, fires, looting, fighting and killing. Civilians were being evacuated from one area to another, houses were being shot at by tanks, cars were being burned by Apache helicopters . . . Baghdad was full of death and destruction on April 9. Seeing tanks in your city, under any circumstances, is perturbing. Seeing foreign tanks in your capital is devastating.

April 9, 2003 was the day Baghdad was declared to be under the control of American troops. The quote is from the weblog of Riverbend, an Iraqi woman, quoted in Gregory, 2004, p. 213.

Case study: Saddam Hussein's use of Arab nationalism and Islam to justify the 1991 Gulf War

The 2003/4 war on Iraq was initiated by the United States within the parameters of the "War on Terror," a key component of its geopolitical code. This war must also be seen as a continuation of an earlier conflict, the Gulf War of 1991, when the US responded to Iraq's 1990 invasion of Kuwait in order to maintain the political status quo in the Middle East and counter Saddam Hussein's attempts to enhance Iraq's power. Looking at Saddam Hussein's use of political rhetoric during the Gulf War provides insights into key ingredients of Arab geopolitical codes as well as the tools used in an attempt to justify the invasion of Kuwait to Iraqis and the wider Arab population.

Underlying Hussein's actions and rhetoric was the politics of Arab nationalism. Arab nationalism can be viewed in two ways. First, as exemplified by Gamal Nasser (President of Egypt 1954–70), Arab nationalism was a political agenda focusing upon Arab unity, a common nation of Arabs that would come together to resist external control by France, Great Britain, and the US, fight the state of Israel, and provide peace and prosperity for the Arab world. As part of this process, Syria and Egypt were united for a brief political moment between 1958 and 1961 when they formed the United Arab Republic. The politics of this expression of Arab nationalism was modern and secular, in opposition to traditional Islamic conservatives (see Khashan, 2000 and Mansfield, 1992 for more on Arab nationalism).

However, the lifespan of the United Arab Republic was brief. It fell apart because of another expression of Arab nationalism, that each Arab state should be an independent sovereign state. The geographic scale of Arab nationalism has, in practice, been centered upon separate national interests. However, as we will see in the rhetorical politics of the 1991 Gulf War, the idea of the unified Arab world was still a backdrop to the politics of the Middle East.

The catalyst for Arab nationalism, in the sense of Arab unity, was the establishment of the state of Israel in 1948, and the consequent wars with Arab states (we will discuss the Arab–Israel conflict in more detail in Chapter 5). Israeli victories in the consequent wars of 1967 and 1973 led to a feeling of Arab humility, and further derailed the unity of Arab nationalism as individual countries made peace with Israel and also allied themselves with the United States, while others proclaimed their anti-Israeli and anti-colonial credentials in the name of Arab nationalism.

Two countries that remained hostile to Israel and the United States were Syria and Iraq. These two countries shared the Ba'athist political ideology, a combination of nationalism and socialism that had its philosophical roots in European left-wing politics. In that sense, it was an imported political ideology that aimed for the secular modernization of the Arab world in order to fight Israel and resist outside "colonial" influence, especially the presence of the United States. As we discuss the geopolitical code of Saddam Hussein it is important to note that at the outset the Ba'athist philosophy was opposed to conservative fundamentalist Islamic political ideologies, but emphasized an Arab path to modern society.

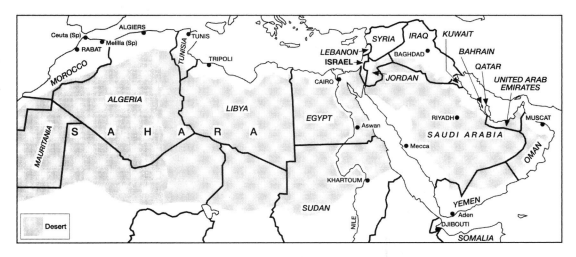

Figure 3.3 The Arab world.

Before the conflict with Kuwait, Iraq's geopolitics had centered upon its conflict with Iran, during which it was supported by the United States of America. After the overthrow of the pro-Western Shah of Iran in 1979 by the fundamentalist Islamic regime of Ayatollah Khomeini, Saddam Hussein's war with his Persian neighbor was seen as a means to prevent Iran expanding its influence in the region. The Iran–Iraq war lasted eight years (1980–8). Estimates of the number of casualties vary, but it is thought that the Iranian dead amounted to between 400,000 and 600,000 and the Iraqis suffered approximately 150,000 casualties (Brogan, 1990, p. 263). The war ended in a stalemate, but the situation was enough for Hussein to claim a "victory" to his people. To maintain his political legitimacy, Hussein emphasized border disputes with Kuwait but used the rhetoric of Arab nationalism.

Iraq and Kuwait had been part of the Ottoman Empire, but Great Britain had established de facto protectorate control over Kuwait before the final Ottoman collapse. At the end of the nineteenth century Britain's policy was to control the Persian Gulf, and though it had no real interest in controlling Kuwait its policy was to exclude other powers. Of special concern to the British was the influence of the Ottoman allies Germany and, to a lesser extent, Russia. What was at issue was the discussion of a railway from Baghdad to the Gulf, which was seen by the British as a means of extending German influence in the region. The key question was where on the Persian Gulf coast would the railway end? To cut a long story short, Britain aimed to determine the location of the railhead by defining Kuwait's borders in such a way as to control the mouth of the Shatt el Arab and make Basra, in Ottoman Iraq, an unattractive choice for the rail terminal (Schofield, 2003).

The view from Iraq was different, however, and remained a viable political stance for Iraqis throughout the twentieth century. In Hussein's interpretation, Kuwait was part of the Ottoman province under the authority of Basra. In this logic, Kuwait should be part of Iraq as Kuwait had been part of the Ottoman territory controlled by what is now an Iraqi city. Both the logic and the historical interpretation were false. Ottoman control of

Kuwait had not really ever existed, and it was outmaneuvered in its political claims to the territory by the British. However, historical record was not the concern of Saddam Hussein; his targeting of Kuwait rested more upon his interpretation of a contemporary material concern: oil (Schofield, 2003 and Slot, 2003).

Kuwait possesses the fourth-largest oil reserves in the world, behind Saudi Arabia, Iraq and Iran (World Oil, 2003). In the rhetoric of Hussein, Kuwait "owed" Iraq because of the service it had performed in fighting a war for all Arabs against Iran. With $40 million of debt from the war, Hussein was looking for political glory by receiving thanks for his efforts from his Arab brothers, but also a fraternal injection of cash to his war ravaged economy. Neither was forthcoming. And so Hussein portrayed the Kuwaiti regime as misusing their oil fortune. Rather than sharing oil wealth amongst the Arab nations, in the form of a regional sharing of wealth, the Kuwaiti elite were, according to Hussein, taking the money out of Kuwait and spending it in an immoral manner. By invading Kuwait, Hussein asserted that he would be claiming Kuwaiti oil reserves for the benefits of all Arabs. In reality, if the invasion had been successful Iraq could have dominated the world's oil production.

The material aspect of Hussein's geopolitical code was domination of the Persian Gulf oil reserves in order to counter the costs of the Iran–Iraq war and provide the economic basis for dominance in the region. These actions were justified, with special focus on the whole Arab world, by framing material concerns within Arab nationalist rhetoric and language of religious identity and conflict in the name of Islamic brotherhood.

One ingredient of Hussein's message was the ideal of Arab unity in which oil wealth was portrayed as a common resource being stolen by the wealthy and immoral emirs (Long, 2004, p. 29). Hussein claimed he would facilitate an Arab nation that "will return to its rightful position only through real struggle and holy war to place the wealth of the nation in the service of its noble objectives" (quoted in Long, 2004, p. 29). In a clever turn of phrase, this was an internal Arab conflict, a *thawra* (revolution) against *tharwa* (wealth) (2004, p. 29).

President George H. W. Bush's efforts to establish a UN-approved, but US-led, coalition to oust Iraq from Kuwait provided another rhetorical angle for Hussein. As US troops were stationed in Saudi Arabia, Hussein began to portray the conflict as a religious one: Islam was under threat from Zionism and colonialism. This clever move by Hussein added the plight of the Palestinians into his message, painting a picture of his geopolitical code as a defense of the most downtrodden of all Arabs, rather than a grab of oil reserves. Also, it turned the coming conflict with US forces into a defense of Islam against a new "crusade," with, of course, Hussein and the Iraqi people in the vanguard. Hussein was portraying Iraqi geopolitics in terms of pan-Arab social justice, the long-awaited stand against Zionism, and a defense of the Islamic religion.

On December 10 the Iraqi Ba'athist party released the following communiqué, which leaned heavily toward the secular language of Arab nationalism:

> Masses of our militant Arab masses, the Arab Socialist Ba'ath Party, which considers this pan-Arab confrontation the Arab nation's battle, calls on you to entrench military cohesion . . . [and] to assert the reality of the dialectical relationship between Iraq's steadfastness and the inevitability of its victory in

بريشة الفنان محمود حمد

Figure 3.4 Saddam Hussein: benefactor of the Arab world.

the crucial battle on the one hand and the intifadah of the occupied territory and
the liberation of Palestine . . . on the other.

(Quoted in Long, 2004, p. 92)

On August 24, Baghdad Radio gave a special broadcast entitled "Muslim Unity
Needed" in which the following geopolitical claims were made:

From the religious point of view a Muslim cannot ask for help from a non-
Muslim under any circumstances but the infidel Saudi ruler asked for the
protection of Israel and the US which are both not only non-Muslim but also
hold a grudge against Islam. It is the same old imperialist ways of intervention
in an occupation of others' territories. The world cannot forget the American

crimes in Hiroshima and Nagasaki nor can the world forget the lesson taught to the Americans in Saigon, Vietnam, and Korea where the Americans' poison gas and napalm proved to be of no effect.

(Quoted in Long, 2004, p.106)

Iraq's forces were defeated overwhelmingly by the number and technical superiority of US-led forces. While a ticker tape parade welcomed home the commander General Schwarzkopf, coalition forces were slow in reacting to Hussein's brutal efforts to cling on to his authority. Uprisings in Shia areas in the south and Kurdish regions in the north were suppressed brutally. In the wake of disturbing pictures of Kurdish refugees, the coalition eventually acted through the imposition of "no-fly zones" that constrained the geographic reach of Hussein's regime.

Part of Hussein's rhetoric in the first Gulf War focused upon his ability to fire missiles, with chemical warheads, into Israel. A few missiles were fired and fell on Israel, but they contained no poison, provided no real threat, and mainly served an ideological goal in illustrating Hussein's role as liberator of the Palestinians. Memory of those weapons loomed large in interpretation of Hussein's denial of UN weapons inspectors. In what is now known to be partial and faulty intelligence, the US government, backed by Great Britain, claimed that Hussein was building a chemical and, perhaps, biological arsenal of weapons that would threaten Israel and US troops in the region. War came again, and this time Hussein was overthrown.

The dynamism of geopolitical codes

Over the years, the geopolitical code of a country will often retain some of its key ingredients. Japan has identified itself as the Pacific ally of the US since the end of World War II, for example, and since 1979 Iran has pursued a code challenging the US in the Middle East. However, geopolitical codes are not completely static. They may change dramatically, as with the end of the Cold War with the collapse of the Soviet Union. Or the change may be more protracted and contested, as with the slow development of what became known in the 1980s as a European Security and Defense Identity by the countries of the EU. In a context such as the Cold War we should expect stability in the geopolitical codes of the main participants, such as the US and the Soviet Union. The situation is likely to be quite different in the post-Cold War world in which the source and nature of threats became more diverse. As a result, representations of geopolitical codes are likely to change too.

Tracking dynamism in a geopolitical code requires a regular event that may be analyzed. For the United States the annual State of the Union speech given by the President allows for such an analysis. This example also illustrates that it is not just the popular media that provides geopolitical representations: governments are also constantly justifying their geopolitical actions. By tracking the use of language in the State of the Union speeches from 1988 to 2008, or the final years of President Reagan's term through the administration of President George W. Bush, we can see how the geographical orientation of the US's geopolitical code changed, as did the tone of the message (Flint *et al.*, 2009). In this way the changes in the US geopolitical code can be seen to be a product of the combination of the will and beliefs of whoever is in the White House on the one hand, and the changing global geopolitical context on the other hand. The technique used was to simply count the number of mentions countries were given within the State of the Union speech. Countries that were mentioned most were, arguably, perceived as having the greatest strategic importance, either as threats or as allies.

Figure 3.5 maps the geographic focus of President Reagan's 1988 speech. Surprisingly it is quite sparse and reflects concern with the Soviet Union and conflicts in Central America. In comparison, an aggregate of the speeches made by President George H. W. Bush (Figure 3.6) shows retention of Reagan's geographical foci and the addition of China and the Middle East. The speeches in President Clinton's last term (Figure 3.7) became more global, to include much of South, Southeast, and East Asia, though after the first Gulf War, prosecuted by his predecessor, there was less focus on the Middle East. The final map displays the global focus of the speeches of President George W. Bush's last term (Figure 3.8). The Global War on Terror was at its height and all the continents, with the exception of Australia, are mentioned. The focus on the Middle East has returned but the regional concentration has expanded to include Egypt and Sudan. (For further discussion of these maps see Flint *et al.*, 2009.)

The sequence of maps shows that over time the key annual event of the State of the Union has been used by the President to draw the attention of policymakers and the public to different parts of the world and the US's engagement with them. In some ways this changing geography is a result of events, such as President George H. W. Bush's focus on the Middle East after Iraq's invasion of Kuwait. In other ways, the changing pattern is a result of different Presidents having the power to change the content of the US's geopolitical code and engage (positively or negatively) with particular countries. In other words, the agency of the President interacts with a changing global geopolitical context to define the content of the geopolitical code.

One thing that a President has more control over is the tone of the speech. Figure 3.9 displays analysis of the speeches to see if the language used focused more upon allies or enemies. A clear pattern emerges, with the two Bush Presidencies predominantly using language identifying threats or enemies and President Clinton's speeches using terminology referring to friends or allies. Though this still may be a function of context to some degree, it is also a function of the tenor created by the speech writer. Is the overarching image one of a dangerous world toward which the US must respond militarily or is it a world of actual and potential allies that are seeking cooperation with the US?

As we saw in the case study of Saddam Hussein and in the analysis of US State of the Union speeches, geopolitical codes can be dynamic. When they are, governmental leaders are faced with the task of justifying such changes to the public. In Hussein's case this was a simple rhetorical switch from nationalist to religious language. In the case of the US, with its global geopolitical code, this requires drawing attention to engagement with different parts of the world. Such engagement is often supported by the language of popular media, as we saw in our discussion of *Reader's Digest* and the Bourne films.

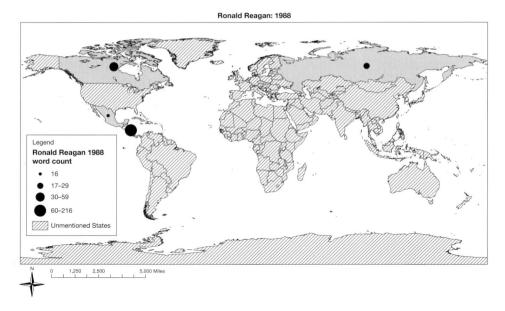

Figure 3.5 The geography of the State of the Union speeches of President Ronald Reagan

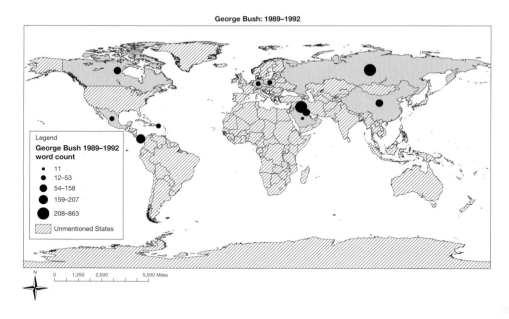

Figure 3.6 The geography of the State of the Union speeches of President George H. W. Bush

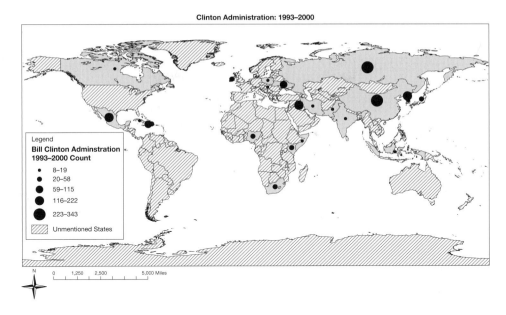

Figure 3.7 The geography of the State of the Union speeches of President William J. Clinton

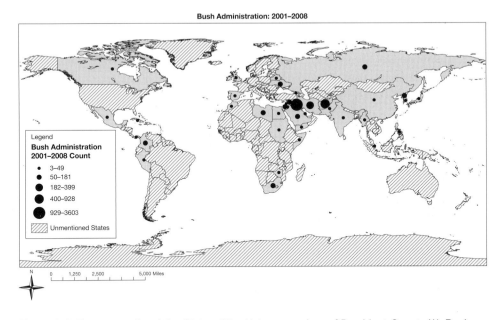

Figure 3.8 The geography of the State of the Union speeches of President George W. Bush

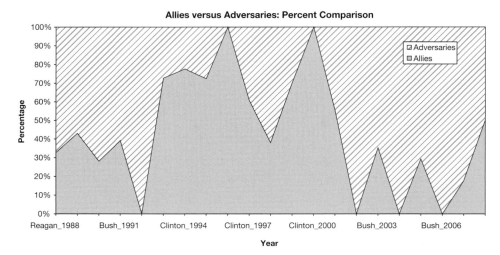

Figure 3.9 The changing emphasis upon allies and enemies in US State of the Union speeches

Summary and segue

In the conclusion to this chapter it is important to emphasize that we have discussed representations rather than "facts." If the calculations for war can be traced to material interests, such as access to oil, governments must usually emphasize values or ideas in justifying their foreign policy, especially when it involves invading a country rather than defending one's own. Two important audiences must be addressed to justify a country's geopolitical codes: the domestic and the international audience. The US, with its global geopolitical code, has a particular burden when it comes to representing its geopolitical practices: it must convince the whole world that it is acting for the benefit of all rather than for its own interests and gain.

The representations of geopolitical codes will use "common sense" or Gramscian understandings of power relations to tell a story that is familiar and appealing to most people. This will likely include dominant ideas of male and female roles, as well as constructed common ideas of places and regions. For the geopolitical codes of countries, another "common sense" understanding is crucial: countries are the obvious, or even "natural," geopolitical agents. For this to resonate with the public requires a sense of nationalism, and it is to that topic we now turn in the next chapter.

Having read this chapter you will be able to:

- Understand that popular culture is an integral part of geopolitics
- Critically evaluate government statements justifying foreign policy
- Relate the use of different justifications used by countries to the different geopolitical situations
- Think critically about the way other countries are represented in popular culture
- Think critically about the way the foreign policy of one's own country is portrayed in popular culture and government statements

Further reading

Dittmer, J. (2010) *Popular Culture, Geopolitics, and Identity*, Lanham, MD: Rowman and Littlefield.

A discussion of the concepts behind the connections between geopolitics and popular culture illustrated by compelling examples and case studies.

Hedges, C. (2003) *War Is a Force That Gives Us Meaning*, New York: Anchor Books.

An exploration, sometimes disturbing, into the way that our individual and collective identities are inseparable from the practice of warfare.

Said, E. (1979) *Orientalism*, New York: Vintage Books.

The seminal text on the representation of "others" in the popular media.

The following readings provide examples of how the representational strategies of geopolitics have been studied:

See the special issue of the journal *Geopolitics* 10 (Summer 2005) on geopolitics and the cinema edited by Marcus Power and Andrew Crampton.

Flint, C. and Falah, G.-W. (2004) "How the United States Justified Its War on Terrorism: Prime Morality and the Construction of a 'Just War'," *Third World Quarterly* 25: 1379–9.

Gregory, D. (2004) *The Colonial Present*, Oxford: Blackwell.

Sharp, J. (2000) *Condensing the Cold War:* Reader's Digest *and American Identity*, Minneapolis: University of Minnesota Press.

4

EMBEDDING GEOPOLITICS WITHIN NATIONAL IDENTITY

In this chapter we will:

- Define the terms nation, state, and nation-state
- Discuss nationalism and its different manifestations
- Discuss the way gender roles are implicated in nationalism
- Make connections between gender roles, nationalism, and geopolitical codes
- Provide a classification of different nationalisms and their impact upon geopolitical codes
- Provide brief case studies of two ongoing nationalist conflicts: Chechnya and Myanmar/Burma
- Discuss nationalism in the context of globalization

The nation: an essential part of geopolitical practice and representation

The previous two chapters have introduced us to geopolitical action or practices (through the concept of geopolitical codes) and the representation of those practices. In this chapter we will try to further understand these key elements of geopolitics by discussing what is commonly called "the nation," is inaccurately called "the nation-state," and should actually be called "the state." The nation has been identified as the key actor in geopolitics. Classical geopolitics was written from a particular national position with the explicit goal of furthering that nation's fortunes, often at the expense or in direct conflict with other nations. Feminist geopolitics has called on us to rethink the us/them and inside/outside binaries that have dominated geopolitical thinking (Giles and Hyndman, 2004), and we will show the benefits and necessity of this idea later in this chapter.

Our first task is to understand why the nation is such an important element of geopolitics and, despite the imperatives of globalization and news ways of seeing geopolitics, why it will remain a key element of geopolitical practice and representation. The connection lies in the way that we have been socialized to think of the state as the nation. In the next section we will explain how and why the state has come to be understood as the nation. For now we must recognize that states (or countries) have been the dominant geopolitical actors. They possess legitimacy to act within their own territory (through the notion of sovereignty or ultimate authority) and are mutually recognized by other states (through international law and institutions such as the UN). States have, literally, shaped our world by fragmenting the globe into territorial jurisdictions so that the world political map is composed of state-ruled and exclusive territories. We look to states to conduct legitimate military actions, regulate multi-national companies, enact and police international agreements on "global" issues such as whaling and climate change, and provide a sense of governance of issues such as nuclear weapons proliferation through the UN. States have been, and remain, the essential and dominant geopolitical actor.

But wait a minute. We started talking about nations and the previous paragraph is all about states. Why? As the next section will show, everyday geopolitical representations encourage us to call states "nations." This is because the state is a geopolitical structure that requires the loyalty and participation of individuals (as citizens) and social groups (such as political parties, religious groups, etc.) to function. The idea of the nation is a more effective representation to generate individual and group identity and loyalty. Humans are apt to die and fight for their nation rather than their state. Hence, we "see" geopolitics through representations that are from a particular national position. A sense of national identity is the building block of geopolitical representations that are dominated by messages of national affiliation and history, and often by threats to that nation from other nations. Or that should be states, right? Yes. And so now that we have seen that nations/states are essential geopolitical actors and play a key role in geopolitical representations we will see how and why we are socialized to use the term "nation" instead of "state."

(Misused) terminology

It is important to get our terminology correct, before we proceed. Up to now, I have mainly used the term "country" when referring to the United States, Great Britain, Iraq, etc. The more precise term is "state." This can be confusing, especially in the US, where the term "state" is used to refer to the fifty separate entities that comprise the country. While discussing geopolitics, however, it is more precise to refer to countries as states. Hence, the United States is actually a state, as are Great Britain, Kuwait, France, Nigeria, etc.

States are defined by their possession of sovereignty over a territory and its people. States are the primary political units of the international system. A state is the expression of government control over a piece of territory and its people. The geographic scope of the governmental control exists in a series of nested scales. For example, the London Borough of Hackney is a scale of government, nested within the Greater London Council, the United Kingdom, and the European Union. In another example, the Borough of Queens is a scale of government within New York City, New York State, and the United States of America. We would refer to Hackney and Queens as the local state or local government. In this book "state" means the country: for more discussion on the political geography of states see Cox (2002) and Painter (1995).

Box 4.1 Processes of state formation

The tradition of geopolitics has focused as much on how states have been created as it has on competition between states. The process of state formation is one in which a single political authority claims sovereignty (or undisputed power) over a clearly demarcated territory. The process involves political centralization and a means of collecting taxes that support the new government institutions. The creation of what we now understand to be a modern state involves eliminating (sometimes by force and sometimes by incorporation into state institutions) regional seats of power and religious claims to authority that would challenge the state. In some cases this may result in mass violence or genocide, as those seemingly threatening or foreign to the state are literally eliminated. Often the process is more political, as some regional autonomy is given to sub-national groups or religious institutions become "nationalized." In the terms of sociologist Michael Mann (1986), the state develops despotic power (or the right and ability to use force within its boundaries) as well as infrastructural power (or tying the country together through the provision of services – roads, education, and a state bureaucracy). Part of this process requires creating a sense of national identity for the whole population so that regional and ethnic identities are subsumed within an overarching national allegiance.

State formation is not just an internal or domestic process. It also has an international aspect. The key historic moment was the Treaty of Westphalia in 1648 in Europe that established states as legitimate and territorial entities and required mutual recognition and interaction between states. States only become fully fledged sovereign states when they are "recognized" as such by other states. Today, such recognition is formalized by a state being part of the UN. Now the whole globe (with the exception of Antarctica) is made up of a mosaic of states. But this is a relatively new situation. It was only in the decades after World War II that swathes of the globe ceased to be under colonial control by European powers and Japan. A wave of decolonization established the map of states with which we are now familiar. In 1945 the UN had 51 original members. Decolonization through the 1950s and 1960s resulted in a new total of 132 members in 1971. Another wave of new member states occurred in the early 1990s, with the collapse of the Soviet Union, leading to states such as Lithuania, Estonia, and Latvia reclaiming their status as independent states and the establishment of states in Central Asia (e.g. Kazakhstan and Turkmenistan): the UN had 184 members in 1993.

The diffusion of modern states from their original formation in Europe across the globe has been categorized as a process of "Europeanization." The model of how a state operates and what institutions are necessary to manage a state certainly has followed the original pattern established in Europe. For example, Japanese modernization in the late 1800s and early 1900s looked to Prussia as a model of what a modern state should be. Many modern states are still strongly influenced by the institutions that were imposed upon them during colonial rule (India and

(Continued)

Australia are good examples). This does not mean that all states are the same. Though all states are centralized, bureaucratic, and territorial entities, the balance of the despotic and infrastructural power varies from state to state (Kuus and Agnew, 2008). A dictatorship such as North Korea is an example of a state where despotic power predominates, while democracies operate through the exercise of infrastructural or integrative power. But what of the contemporary developments that were sparked by protests and uprisings in Arab countries in 2011? At the time of writing it is impossible to say anything definitive. However, the desire of most protestors appears to be for a form of liberal democracy that is situated within the existing model of a modern state evolved from the Western model. Claims for a new type of state based on a restricted reading of Islamic texts seem to be marginalized.

The terminology may be confusing because it is so widely misused. Instead of using the term state, the term nation or national is usually substituted. Hence, we stand for the national anthem, rather than the state anthem; in the World Cup and the Olympic Games we say that national teams compete, rather than state teams. However, the term nation has a very specific meaning that, if we focus on the definition, should not be used in this way. A nation is a group of people who believe that they consist of a single "people" based upon historical and cultural criteria, such as a shared language. In some contexts, membership of a nation will be granted only if inheritance, or blood ties, to members of a particular group can be established, but most nations do not require such blood ties. An introductory discussion of nations and nationalism can be found through the work of A. D. Smith (1993 and 1995).

The geopolitics of nationalism I: constructing a national identity

Dulce et decorum est pro patria mori? (See Box 3.1.) Can people be motivated to kill and die for a government bureaucracy? It is hardly a sense of attachment to the Ministry of Defense or the State Department that inspires people to fight. Instead, the ideology of nationalism has equated national well-being with control of a state, the state and nation become synonymous, and the sense of identity is focused upon the nation rather than the state. Nationalism is the belief that every nation has a right to a state, and, therefore, control of a piece of territory. The ideology of nationalism claims that a nation is not fulfilled, the geopolitical situation is perceived to be unjust, if a nation does not have its own state. The geopolitics of nationalism has resulted in millions of deaths, as people fought to establish a state for their nation, and defend their states, in the name of national defense, against threats, real and perceived.

The state is equated with the nation through another term, the "nation-state": the notion that each state contains one nation. Hence, the Australian nation-state, for example,

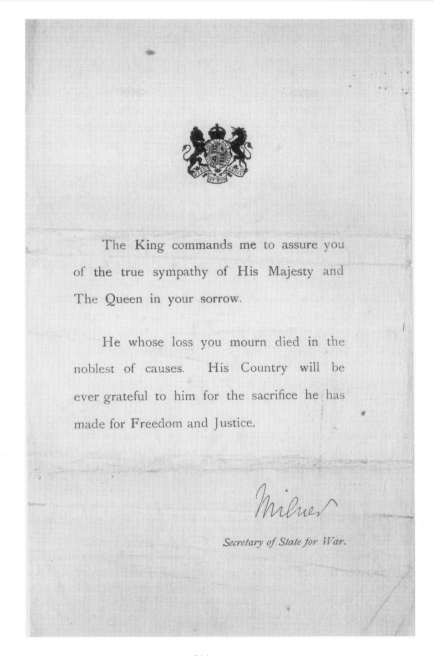

The King commands me to assure you of the true sympathy of His Majesty and The Queen in your sorrow.

He whose loss you mourn died in the noblest of causes. His Country will be ever grateful to him for the sacrifice he has made for Freedom and Justice.

Milner

Secretary of State for War.

Figure 4.1 World War I telegram to next of kin.

refers to an Australian nation contained within the Australian state. Such is the ideology. The reality is much different, and the potential for conflict is large. Nearly all states have a diverse population of cultural groups, some of which may define themselves as separate nations (Gurr, 2000). In some situations, a national identity may take precedence over an ethnic identity (Arab-Americans or Italian-Americans, for example). In other cases, a group may demand a degree of autonomy, especially in terms of cultural practices such as the use of language in schools. When a cultural group defines itself as a nation, often there are demands for a separate state for that nation, the politics of nationalism. We will look at the politics of creating nation-states in two ways: top-down and bottom-up.

Top-down nationalism refers to the role of the state in creating a sense of a singular, unified national identity (Mosse, 1975). The United States is, perhaps, the best example of this process. The history of the United States defines its national identity as an immigrant nation: a collection of individuals from national groups across the globe. The practice of the state has been to ensure a centripetal political force: that such a collection of nations does not create conflict, but is "a more perfect union," an American nation. Education is the vehicle for this continual process of creating a nation-state. Children pledge allegiance to "the flag" at the beginning of each school day. The American nation is celebrated in song, dance, and study – the mythology of the nation, the sense of unity, and the child's place within it is created in a "banal" or everyday manner (Billig, 1995). The celebration of the American nation (and also the Australian and Canadian national histories) illustrates that there are positive interpretations of nationalism as a collective identity that transcends ethnic differences.

Ironically, the dominant mythology of the United States as an immigrant land of opportunity rests upon a history in which different cultural groups have suffered at the hands of the state: racist immigration policies targeting Chinese, such as the Chinese Exclusion Act of 1882, the near-genocidal Indian Wars of the 1800s, as well as the enslavement of black Africans and the African-Americans' struggle for civil rights that continues today, and the contemporary harassment of Arab-Americans at airports and other security points in the name of the War on Terror. However, the power of the United States' national identity is that despite the discrimination that successive waves of immigrants have experienced, and still do, the desire to be part of the American nation remains strong, and the degree of assimilation is high, compared to other countries.

The top-down nationalism of the United States illustrates the way the state apparatus has been brought to bear to create a nation. It is a form of nationalism; it promotes the ideology that the state is the natural and obvious political geographic expression of a singular nation. Funnily enough, it is not the type of politics we usually think about when we label a politician a "nationalist": such terminology is usually a form of epithet referred to "monsters" such as Slobodan Milosevic or Adolf Hitler, for example.

Activity

How and where did you learn your national history? Think of the settings (home, school, etc.) where you were exposed to this history and the form the history took (books, films, lessons, etc.). Write down two or three of the key ingredients of the history and what "moral" or story they may say about the particular national character. Write down two or three key historic events that are usually ignored or played down in the national history of your country. How do these less-discussed events contradict the portrayal of the national character you identified from the dominant narrative?

The geopolitics of nationalism II: the process of "ethnic cleansing"

Let us now turn to the politics of violent nationalism, or bottom-up nationalism; the type of nationalism that makes the headlines (Dahlman, 2005). Nationalism, in this sense, is the goal to create a "pure" nation-state, in which one, and only one, culture or national group exists. This geopolitical perspective views a nation-state as somehow tainted, weak, a geopolitical anomaly, if it contains multiple nations or ethnicities. Instead of the politics of assimilation, the geopolitics here is of expulsion, and eradication. Bottom-up nationalism is what has become known, almost nonchalantly, as "ethnic cleansing." Though it is the bloody actions of "ethnic cleansing" (the killing and rape) that is the "sharp end" of this form of nationalism, the way the term has become readily, and quite uncritically, adopted by mainstream media as a handy phrase to "make sense" of an event also shows that we are implicated too. As viewers, the pervasive ideology of nationalism makes the goals of "ethnic cleansing" understandable: it is, simply, the most extreme form of the politics of exclusion that underlies discussion of immigration and refugee policies in "civilized" debates in the British parliament, for example. The politics of otherness related to particular territories is the underlying geopolitics.

The process of "ethnic cleansing" can be illustrated schematically. In the first diagram (Figure 4.2), two neighboring states are both multi-national: Triangle State is populated mostly by people ▲ with a scattering of people ● near the border, and Circle State displays the opposite pattern. Also, there are both ▲ and ● people living outside the borders of these two states. The existence of people of different nations does not determine conflict; most states exhibit this mixture of nations without violence. But, in some cases, politicians gain prominence on the back of calls to alter the multi-national make-up of the state, often by blaming economic and social woes upon the presence of a minority nation.

The drive to create a "pure" nation-state is illustrated in Figure 4.3. In what has become known as "ethnic cleansing" the minority ● nation is expelled from the landscape of the Triangle State. Expulsion usually consists of violence against people and their property that forces them to flee for their safety, leaving their property and possessions behind. Expulsion usually takes place in conjunction with eradication (Figure 4.4): the slaughter usually of young men and the rape of women to prevent the reproduction of future generations. Rape is a powerful weapon (see Chapter 9). Women are "defiled" in order to "pollute" the purity of the nation. In our schematic example, Circle State has retaliated to the expulsion of people ● in Triangle State by killing people ▲ within the borders of the Circle State. The goal of ● and ▲ nationalists is the creation of a state containing one, and only one, national group.

Sometimes, the geopolitics of nationalism will stop here. In other cases, there may be a further violent step (Figure 4.5). After the bout of ethnic cleansing, Circle State may be pure, but for some the process is incomplete: not all of the members of nation ● reside within the borders of Circle State. In a fundamentalist interpretation of nationalism, the members of nation ● in Triangle State are unfulfilled, denied their right to participation in the ● nation-state. The result is the mobilization of force to change the borders of Circle State so that all members of nation ● now reside within it. At the same time, any

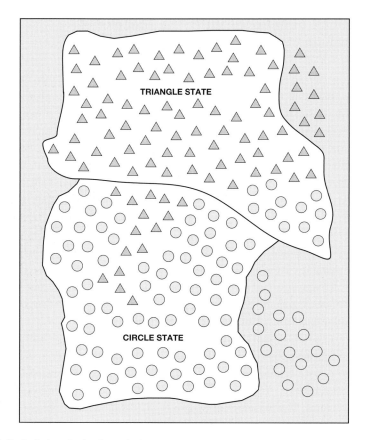

Figure 4.2 Prelude to ethnic cleansing.

▲ people in Circle State who survived the killing flee or are expelled. The "purity" and "wholeness" of nation ● has been achieved; a lot of blood has been spilt.

Case study: Chechnya

Conflict in the breakaway Russian province of Chechnya has been an ongoing and brutal process of nationalist geopolitics (BBC, 2004; Knezys and Sedlickas, 1999). The struggle is about, on the one hand, the territorial integrity of Russia and, on the other hand, the desire by Chechens for their independent country. In other words, there is a fight over the establishment of Chechnya as an independent nation-state. The population of Chechnya is approximately one million and composed primarily of Chechens and Ingushes as the dominant ethnic groups in the region. Most Chechens are Sunni Muslims. Chechen and Russian civilians alike have suffered in the conflict. Approximately 200,000–250,000 Chechens have been forced to leave their homes, most of them during the Russian invasion of 1999, while Russian civilians have had to live in fear of suicide bombings and various other forms of terrorist attacks throughout Russia. For more information on the conflict,

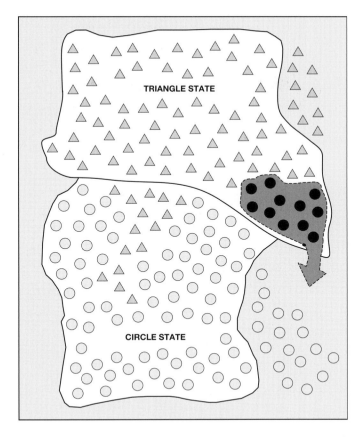

Figure 4.3 Ethnic cleansing: expulsion.

refer to the readings listed at the end of this chapter, which are also the sources of the information below. The purposes of this case study are:

- to provide historical background to help you interpret a recurrent geopolitical conflict;
- to illustrate how nationalist conflicts are viewed differently by different social groups;
- to offer a concrete example of the concepts discussed in the chapter.

History of the conflict

1893: While industrialization is sweeping over Russia, oil is discovered in Chechnya (at this time a part of Russia) – the area becomes increasingly important to Russia.

1890s: Russia builds the Vladikaukaz railroad line through Chechnya – Chechnya is a key route to southern regions of Russia.

1914: By this time, Chechen oil comprises 14 percent of Russia's oil production.

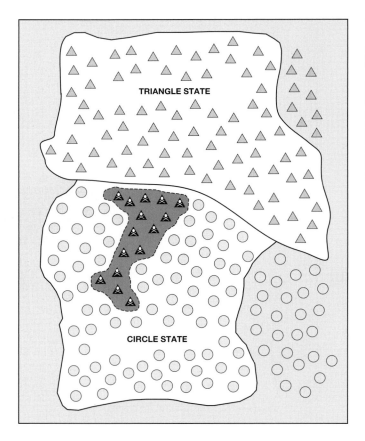

Figure 4.4 Ethnic cleansing: eradication.

1917: Beginning of the Bolshevik Revolution – Czar Nicholas II and then the Provisional Government are both ousted by the Bolshevik Party.

1918: The Mountain Republic is established, only to be taken back by Lenin in 1921.

1922: The Mountain Republic is officially dissolved into the Chechen Autonomous Oblast on November 30.

1923: Lenin's Congress officially adopts the policy of *korenizatsiya* (indigenization), encouraging different nations to use their languages and celebrate their cultures and instilling a sense of ethnic and national awareness in minority groups.

1934: The Chechen Autonomous Oblast is merged with the Ingush Autonomous Oblast.

1936: The combined oblast rises to the status of ASSR (Autonomous Soviet Socialist Republic).

1944: Stalin begins deporting Chechen and Ingush people to Siberia and Central Asia, accusing them of conspiring with Nazis.
  • "On February 23, 1944 over 500,000 Chechens and Ingush were transported to northern Kazakhstan for an exile that lasted 13 years" (German, 2003, p. 4). Chechen language publications were banned and the term "Chechen" [plus

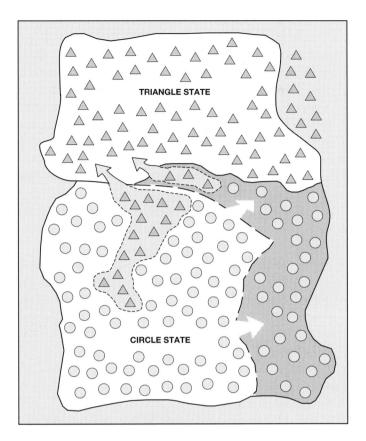

Figure 4.5 Ethnic cleansing: expansion.

descriptive terms of other nationalities that had been deported] was erased from Soviet textbooks and encyclopedias.

1957: January – Nikita Khruschev reestablishes the Chechen-Ingush Autonomous Soviet Socialist Republic.
- "The return of the deportees did aggravate tensions in this region, which could not support an influx of people who were without homes or employment" (German, 2003, p. 19).

1985: Gorbachev rises to power.
- "The advent of glasnost and perestroika [the political and social changes that signaled the collapse of the Communist Party's control of the Soviet Union], and consequent relaxation of previous restrictions, heralded the appearance of popular fronts demanding greater autonomy for the manifold ethnic groups" (German, 2003, p. 14).

1990: November Former Communist leaders "convoked a Chechen National Congress . . . and invited the recently promoted General Dudaev – who had never lived in Chechnya – to head the nationalist movement" (Evangelista, 2002, p. 16).

1991: The collapse of the Soviet Union.

- November 1: Dzhokhar Dudayev, after winning a presidential poll, proclaims Chechnya independent of Russia.
- Russian President Yeltsin declares Martial Law in Chechnya and Ingushia, sending 1,000 Internal Affairs Ministry troops – but they leave without ever disembarking from their aircrafts as crowds block the airport.
- Chechnya begins to develop its army. It is to consist of 11,000–12,000 troops with approximately 40 tanks and 50 units of various armored equipment.

1992: Chechnya adopts a Constitution.

1994: Russian troops invade Chechnya to end the independence movement.
During the twenty-month war that follows, approximately 100,000 people, many of them civilians, are killed.

1995: Chechen rebels seize hundreds of hostages at a hospital in Budennovsk, southern Russia. Over 100 are killed in the initial raid and the subsequent unsuccessful Russian commando operation.

1996: • April: President of Chechnya, Dudayev, is killed by a Russian missile attack. Succeeded by Zemlikhan Yandarbiyev
- May: Yeltsin signs short-lived peace agreement with Yandarbiyev.
- August: Chechen rebels attack Grozny (rebel chief of staff Asian Maskhadov and Yeltsin's security chief Alexander Lebed sign the Khasavyurt Accords – ceasefire).
- November: agreement of Russian troops' withdrawals.

1997: January – Aslan Maskhadov wins Chechen presidential elections. His presidency is recognized by Russia.

1999: • January: Maskhadov announces his three-year plan to phase in Islamic shariah law.
- March: Moscow's top envoy to Chechnya is declared missing (his body is found in Chechnya one year later).
- July/August: Chechen insurgents begin crossing the border into Dagestan to assist the overthrow of the Russian government and the establishment of a separate state. Maskhadov tried to maintain friendly relations with Russia and appeals to Chechens to leave Dagestan.
- September: a series of bombings targeting Russian civilians, attributed to Chechen rebels, provokes the Russian government to take action.
  – Approximately 300 Russian civilians killed.
- October: Russian government launches assault into Chechnya and recaptures the breakaway region of Dagestan.
  – It is estimated that approximately 200,000 refugees flee Chechnya for neighboring Russian republics.

2000: • Russian troops capture Grozny. Russia declares rule of Chechnya. War continues in mountainous areas.
- June: Akhmat Kadyrov is named the head of administration in Chechnya by the Russian government.

2001: • January: rumors of human rights atrocities begin to circulate in the Chechen village of Dachny.

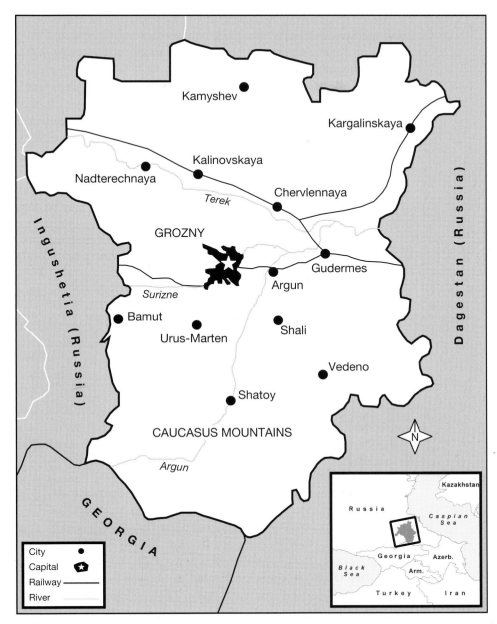

Figure 4.6 Chechnya.

- February: the first body is found in a mass grave in Dachny.
  - After later inspection, it is reported that forty-eight bodies have been found, thirty-four of which are never to be identified.
- 2002: • October: Chechens seize an 800-person theater in Moscow.
  - One man is shot and killed by the hostage-takers.
  - 118 civilians (along with fifty hostage-takers) die as a result of gas used by the Russian government to flush out the terrorists.

2003:   • March: Chechens vote in referendum that makes Chechnya a separatist republic within Russia. Multiple suicide bombings occur following the referendum.
        • Akhmat Kadyrov becomes the official president (many groups question the legitimacy of the elections).

2004:   • February: former Chechen president, Zelimkhan Yandarbiyev, is murdered in a car bombing in Qatar.
          – Two Russian intelligence officers are sentenced to life in prison after admitting that the attack was ordered by the Russian government.
        • 9 May: Akmat Kadyrov and five others are killed in a bombing in a stadium in Grozny. Fifty-six are wounded in the attack.
          – "The blast comes a few weeks after Putin, in his annual state of the nation address, proclaimed the 'military phase of the conflict may be considered closed' in Chechnya" (CNN, 2004).
          – Warlord, Shamil Basayev, declares responsibility for the attack.
        • The conflict spreads into neighboring Ingushetia – known to be a peaceful safe-haven.
          – "A year ago it was an idyllic farming region but now many locals refer to it as 'a second Chechnya'" (Walsh, 2004).
          – Cassette tape surfaces with a Russian soldier's confession to the kidnapping of a senior aide to the top prosecutor of Ingushetia – this aide was pursuing prosecutions against Russian soldiers who were accused of abuses.

2005:   • Ramzan Kadyrov (son of Akhmad Kadyrov) is appointed caretaker prime minister.
        • Aslan Maskhadov is assassinated by Russian forces. Maskhadov had led an unrecognized separatist state within Chechnya (Ichkeria) that challenged the Russian regional government. Abdul Khalim Saidullayev becomes Maskhadov's successor.

2006:   Russian forces assassinate Saidullayev, who is succeeded by Doku Umarov.

2007:   Umarov abolishes the Chechen Republic of Ichkeria and proclaims a Caucasian Emirate, with himself as Emir. The Emirate is largely unrecognized.

2009:   • Russia formally ends counter-terrorist campaign and withdraws much of its army.
        • Separatist leaders call for end of armed resistance.

Experiencing the conflict

Anna Politkovskaya was a leading investigative reporter for the Russian newspaper *Novaya Gezeta*. She was murdered for writing stories that exposed the brutality of the Russian campaign in Chechnya. She reported on the conflict, including the plight of refugees, bombed by Russian forces as they fled:

> They fly so low that you can see the gunners' hands and faces . . . They seem to be laughing at us crawling comically down below – heavy old women, young girls, and children.

(Politkovskaya, 2003, p. 33)

For those who remained in Chechnya, they were targets of both the Russian military and Chechen groups who targeted civilians who worked for or cooperated with the Russian-friendly government. Politkovskaya describes the "pits," where the Chechens are kept to await questioning – in the middle of winter without any other shelter:

> It's about ten by ten . . . Despite the frost, there is a distinct odor. That's how they do things here – the Chechens have to go to the bathroom right in the pit.
> (2003, p. 50)

> An old woman, already a grandmother, accused of harboring militants spent twelve days in the pits. She was taken out, tortured by electric shock, never questioned, and handed over to her family and neighbors after they gathered enough money for ransom.
> (2003, p. 48).

Residents of Moscow were hostile to Chechen and Ingushetia refugees, especially after the hostage situation at the Moscow theater in 2002. Human Rights Watch (2004) cites various abuses of ethnic Chechens in Moscow which include: arbitrary identity checks and detention, the planting of drugs or weapons, obstruction of registration in Moscow, harassment of unregistered Chechens in schools, and pressure on landlords to evict Chechen tenants. Human Rights Watch conducted interviews with victims of various forms of discrimination. One woman told them that she tried to register in Moscow but the registration official told her, "Your wanting to register in Moscow is the same as going to the White House and asking to live there" (Human Rights Watch, 2004).

Even though there was popular hostility to the refugees, Russian citizens were unsupportive of the war. In response, Yeltsin's administration tried to justify the invasion of Chechnya by publicizing Dudaev's most sensational statements in order to instill fear in the Russian people. These included "remarks reportedly made to a Turkish journalist during the war that . . . he would personally fly a bomber to Moscow to retaliate" and that he would kill Russian prisoners of war who tried to assist Dudaev's opponents (Evangelista, 2002, p. 35). However, the development of terrorist attacks has resulted in a greater support for the war among Russian civilians. For example, the bombing of an apartment building that housed only civilians, and other terrorist attacks, had "a traumatic and galvanizing effect on Russian public opinion, comparable to what happened in the United States after the attacks of September 11, 2001. They provoked widespread and uncritical support for expanding the war against Chechnya" (2002, p. 67). Other actions that galvanized public support were the December 1996 kidnapping of twenty-two Russian Interior Ministry troops in Chechnya, and, two days later, the kidnapping of a government delegation, on its way to Chechnya for talks with Chechen officials. The same night six medical personnel from the Red Cross were shot and killed south of Grozny. The following day six Russian civilians, living in Grozny, were killed (2002, pp. 46–7).

Russian policy toward Chechnya has been contested, though on the whole the attitude of the political parties was toward a violent end of the conflict to maintain Chechnya as part of Russia. United Russia, a pro-Putin party, hated by Chechen separatists, wishes that Chechnya remain a part of Russia. The Communist Party (of the Soviet Union/of the

Russian Federation) also wanted Chechnya to remain part of Russia. Yabloko, a more liberal democratic party, was very critical of the war but has toned down its criticisms over time (LaFraniere, 2003). The Union of Right Forces, also liberal, was more critical of Moscow's approach to the Chechen conflict (LaFraniere, 2003). The Liberal Democratic Party, an extreme right-wing party, issued violent ultimatums to the Chechens, threatening their annihilation by military force (Gusev, 1996).

What of the sorry situation of the Russian soldiers who were sent to fight in Chechnya? In the Russian military, approximately 3,000 non-combat deaths occur each year: the result of suicide and beatings incurred at the hands of senior-ranking officials. Malnourishment is also common (Associated Press, 2003). Given these awful conditions, parents and draft-age youths do everything that they can in order to avoid the draft. "Thousands of young men avoid the twice-yearly call up by obtaining educational deferments, medical and family exemptions, paying bribes, or simply dodging military authorities" (Associated Press, 2003). The Committee of Soldiers' Mothers was a strong force in ending the harsh conditions for soldiers. They vehemently voiced their opinion against the status quo and lack of sympathy that exists in the Russian government. The Committee did a lot to help parents gain information that had not been readily available in the past. It was involved in "supporting efforts of parents to travel to Chechnya and rescue their sons from the army or at least find out how they died and recover their bodies" (Evangelista, 2002, p. 42).

Popular sentiment in Russia toward the conflict was a combination of animosity toward the Chechens, especially the refugees, but also discomfort over the conduct of the war, and the situation of the draftees. However, this did not stop Vladimir Putin using the US's War on Terror as a vehicle to strengthen his commitment to the conflict and justify its continuation. Putin made links between Chechen terrorists and al-Qaeda, saying that some of the fighters were trained in the same Afghan camps. Prior to the beginning of the War on Terror, the countries of the EU were very critical of Russia's handling of the Chechen issue. On September 26, 2001, on a visit to Germany, Putin was promised by Schroeder that criticism would tone down and that there would be a "differentiated evaluation" of the matter (Evangelista, 2002, p. 180). Russia used this opportunity to convince the rest of the world of its shared values. For example, at a meeting of the world economic leaders in February 2002, Russia's Prime Minister, Mikhail Kasianov, said "Russia understands better than any other nation what has happened in American" (Cullison, 2002).

There are also different views of the conflict from the Chechen perspective. Though a sense of national identity is shared, some, such as the Chechen leader Maskhadov, fought for a moderate form of independence – with a peaceful break from Russia, followed by cooperation with Moscow and other regional neighbors. On the other hand, many Chechens believed in a harsh break from Russia, and the construction of an Islamic state "encompassing Chechnya and Dagestan and perhaps other Muslim peoples of the North Caucasus" (Evangelista, 2002, p. 50). However, all the nationalist visions of Chechnya rested upon a shared and dominant history of injustice at the hands of the Russian government, dating from Stalin's regime. Nationalist sentiments have risen from the continued fight with Russia: "The deportations failed to break Chechen resistance and instead contributed to an abiding attachment to the homeland and a smoldering sense of grievances" (Evangelista, 2002, p. 13).

The conflict in the region has continued and spread into neighboring Dagestan and Ingushetia. Given the role of Russia as ally in the War on Terror, producer of oil and natural gas, and influential power in Central Asia and the Trans-Caucasian region, Chechnya will be a part of the geopolitical calculations of Russia and other countries. Perhaps, as Politkovskaya (2003, p. 29) noted, the tendency will be to ignore the conflict: "Everywhere they invite me to make a speech about 'the situation in Chechnya,' but there are zero results. Only polite Western applause in response to the words: 'Remember, people are continuing to die in Chechnya every day. Including today.'"

Amnesty International continually disseminates reports to governments and the general public highlighting the brutality of Russian actions in Chechnya that are intended to suppress any resurgent nationalist conflict. For all intents and purposes the rule of law has been suspended:

> Efforts have been made by the Russian authorities to create stability and to address the destruction caused by the armed conflict in Chechnya. However, impunity for human rights violations and the absence of the rule of law remain major obstacles to real and lasting stability in the region. The civilian population remains subject to violence from both sides, armed opposition groups as well as law enforcement bodies, and continues to be deprived of access to justice. Torture and ill-treatment, enforced disappearances, indiscriminate killings and arbitrary detention are pervasive. Investigations into human rights abuses are ineffective and flawed, often resulting in impunity for the perpetrators, perpetuating a lack of trust in state institutions and the justice system as a whole. The gagging of civil society, the indifference with regard to the murder of prominent human rights defenders and threats against journalists and human rights activists, which forced many to leave the country or to end their activities aimed at promoting and protecting human rights and ensuring justice, have seriously impacted on the right to freedom of expression.
>
> (Amnesty International, 2010).

The case study of Chechnya illustrates the tenets of nationalism, where individual and group identity is linked to sovereignty over a particular piece of territory. The nationalist claim is based upon a history that includes injustices, and so promoting the "need" for a protective nation-state. But the geopolitics of nationalism gives rise to conflicts within national politics, as is evident in the broad party support for Russia's policy compared to the disaffection of soldiers and their families. Also, the existence of an emergent separatist state within Chechnya illustrates the dynamic ever-evolving nature of nationalism. Noting the role of families in nationalism requires us to pay careful attention to the role of gender in nationalist politics.

Gender, nationalism, and geopolitical codes

In "Dulce et Decorum Est" Wilfred Owen portrays the harsh world of the battlefield, but it is very clearly a masculine world. In World War I, the roles women were to play were mainly limited to exhorting the troops to go off to war. Other classic writing on warfare is also a story of men under fire. In the novels *The Naked and the Dead* by Norman Mailer and *The Thin Red Line* by James Jones, women are "back home," to be returned to and remembered, and their potential infidelities a further source of stress. Homosexuality is rarely a topic, though *The Thin Red Line* is a notable exception in its casual recognition that men found comfort with men during warfare.

The masculine nature of geopolitical codes is the goal of our discussion. To get there it is necessary to explore the gendered nature of nationalism, with special reference to the use of nationalism at times of conflict. We began this chapter with an emphasis on "naming" so that we can begin to understand nationalism as a political process of joining a nation and a state. Naming is not just a matter of academic classification; it also refers to the particular label given to an abstract concept, such as "the state," in order to give it popular meaning and salience. Think of other, more colloquial names for the nation. Some that come to mind are homeland, motherland, and even fatherland. Fatherland is noteworthy because this particular gender reference to the nation has a very negative historical connotation: Nazi Germany, or nationalism gone "too far." Instead, we are more comfortable with thinking of the nation as the "motherland," with its references to nurturing, comfort, and sense of belonging. The nation is, as Anderson (1991) noted, an imagined community: meaning that we think of ourselves as part of a national community, but we will never interact with the vast majority of its members in any meaningful way. Instead, the sense of community is "imagined" through national events such as sporting fixtures, elections, the funerals of statesmen and women, natural disasters, etc.

Perhaps a more accurate description of the nation is an "imagined family," and a patriarchal one at that. For notions of the "motherland" imply a particular role for women; they should be active in the procreation and socialization of the nation's future generations, and their domain is the home. The flip side to this gendered role is that men are seen as the defenders and rulers of the nation – they inhabit the "public spaces" of government, business, and the military. Hence, the dominant narratives of *The Naked and the Dead* and other war novels: the men must fight for and defend their women "back home," but the individual is only complete (and by extension so is the nation) when functioning as a household of a man and a woman: this, the narrative says, is what is to be fought for.

Feminist scholars have shown the gendered division of labor within politics, and foreign policy in particular (see the essays in Staeheli *et al.*, 2004). Despite some positive changes in attitudes toward women, and legal recourse to equality, ideology is harder to change. By ideology I do not just mean the overt sexism of some individuals and political agendas that promote the role of the woman as "homemaker" and or sexual object. Such agendas are by their very brazenness perhaps relatively easy to challenge. More threatening are the insidious or banal practices that promote gender roles that limit women's participation in public space. Nationalism remains an exceptional tool for defining gender roles, and perhaps especially at a time of conflict.

Women and the War on Terror

In the United States after the terrorist attacks of September 11, 2001 the words "hero" and "nation" were pervasive and intertwined. Without dismissing the bravery of firefighters, police officers, and others who lost and risked their lives in the wreckage left by the terrorist attacks, it is also striking how gendered this narrative became. Men were the heroes, women the homebound helpless victims. The term "brotherhood" was used repeatedly in references to the New York Fire Department. The bravery of the "brotherhood" of rescue workers on that day was portrayed as a purely masculine pursuit: barely mentioned were the women rescue workers who also served (Dowler, 2005).

Of course, women did play an important role in the nationalist story that was told in the wake of these terrorist attacks; the tragedy of widowhood was exposed to a nationalist light. Again, one must emphasize that an academic analysis of the victims of 9/11 is not intended to diminish or demean individual loss and suffering. However, it is important to see how an individual's loss becomes part of a national tragedy or episode that, in turn, feeds into the redefinition of a geopolitical code, especially its military role. The following excerpt from President Bush's State of the Union speech of January 29, 2002 is particularly illustrative:

> For many Americans, these four months have brought sorrow, and pain that will never completely go away. Every day a retired firefighter returns to Ground Zero, to feel closer to his two sons who died there. At a memorial in New York, a little boy left his football with a note for his lost father: Dear Daddy, please take this to heaven. I don't want to play football until I can play with you again some day.
>
> Last month, at the grave of her husband, Michael, a CIA officer and Marine who died in Mazur-e-Sharif, Shannon Spann said these words of farewell: "Semper Fi, my love." Shannon is with us tonight. (Applause.)
>
> Shannon, I assure you and all who have lost a loved one that our cause is just, and our country will never forget the debt we owe Michael and all who gave their lives for freedom.
>
> www.whitehouse.gov/news/releases/2002/01/20020129-11.html
> President Delivers State of the Union Address (title of web page
> accessed December 22, 2004)

In this segment of the speech the tragedy suffered by men in the defense of the nation is clearly flagged, in terms of losses suffered "at home" and also "abroad." Shannon Spann's loss was made a public event as television cameras broadcast her acknowledgment of the applause from the country's elected officials. Semper Fi (or Semper Fidelis) is the Latin motto of the United States Marine Corps, and is translated as "always faithful": Faithfulness that is promised to God, country, family, and the Corps. The use of Semper Fi in the State of the Union speech hits upon different aspects of the nationalist story: the women back home are "ever faithful" while the nation's menfolk are away fighting (contrary to the constant worries about fidelity of the characters of war novels); such faithfulness is part of the institution of holy matrimony, blessed by a God that is blessing the fighting too; the country demands a loyalty that requires people to put their lives at risk, despite the claims to faithfulness to family. Loyalty to nation and the Corps takes precedence over loyalty to family.

The point to emphasize here is that the practice of geopolitics requires geopolitical agency in many different settings, including the home. The geopolitical actions of states are dependent upon the actions and sacrifice of people like Michael Spann. How can his wife's loss, or the loss of any individual fighting for any country, be justified? In other words, a foundational ideology, applicable in all countries, is necessary to justify the conflict that is undertaken as part of a geopolitical code. Nationalism plays that role by creating a sense of "community" and allegiance that warrants sacrifice.

Gendered nationalism and the masculinity of geopolitical codes

Though each country's nationalism is unique, in the sense of its particular history, nationalism is still consistent in defining particular gender roles, even during what are described as "revolutionary" situations such as Cuba. Women consistently are identified with a subordinate role and their access to public positions is limited. In this sense, nationalism can be seen as a structure that contains a message of the "proper" roles for men and women. At times of conflict a state's geopolitical code may reinforce these gender roles as warfare intensifies the different expectations of sacrifice expected of men and women: the "heroic" actions of fighting men and the "stoic" sacrifice of the women who are left behind. There is a constant feedback in this relationship too. As men continue to dominate the public sphere, a masculine perception of the world is continued, with the inevitable result of aggressive geopolitical codes and wars.

Nationalism requires the construction of difference between the populations of different states. Such difference allows for the construction of "enemies," "threat," and "danger" as part of a state's geopolitical code. These notions are dependent upon a dominant military view in society that commands a particular vision of "a dangerous world" and how to respond. From a feminist perspective the dominant military view rests upon a masculine view of the world: the implication is that individual gender roles, geopolitical codes, and the structure of global geopolitics are connected in practice and ideology.

The concepts of militarism and militarization are related to how geopolitical codes are constructed and what they contain. The core beliefs of militarism are (Enloe, 2004, p. 219):

(a) that armed force is the ultimate resolver of tensions;
(b) that human nature is prone to conflict;
(c) that having enemies is a natural condition;
(d) that hierarchical relations produce effective action;
(e) that a state without a military is naïve, scarcely modern, and barely legitimate;
(f) that in times of crisis those who are feminine need armed protection;
(g) that in times of crisis any man who refuses to engage in armed violent action is jeopardizing his own status as a manly man.

In an extension of these concepts, Bernazzoli and Flint (2010, p. 159) point out that militarization also requires connecting these ideas to national identity through notions of patriotism and moral right. Hence five other beliefs that underpin militarism should be added:

(h) that soldiers possess certain values and qualities that are desirable in civil society;

(i) that military superiority is a source of national pride;

(j) that those who do not support military actions are unpatriotic;

(k) that those who do not support military actions are anti-soldier;

(l) that for a state to engage in armed conflict is to serve the will of a divine being.

Militarism is, then, an ideology, a particular view or understanding of society and how it should be organized. It is a different ideology than nationalism, but they are usually found hand-in-hand. Related to militarism is militarization, "the multitracked process by which the roots of militarism are driven deep into the soil of society" (Enloe, 2004, pp. 219–20). One of these processes is the way the military is constructed as a masculine institution and war as a masculine enterprise. In this way, the masculine nature of militarization is complemented and enhanced by the gender roles promoted in nationalism, and vice versa.

The implications of militarization are individual, national, and global. In terms of the construction of geopolitical codes, militarization is seen as a foreign policy issue because of the dominant influence of the military in forming codes, and equating security with military matters. Most importantly, militarization is especially successful when civilian policymakers acquiesce to a foreign policy implemented by force (Bacevich, 2005; Enloe, 2004).

The militarization of a geopolitical code rests upon the dominance of men in positions of public office, who are willing to facilitate a foreign policy that rests upon masculine

---

Box 4.2 Masculinity as a foreign policy issue

"The militarization of any country's foreign policy can be measured by monitoring the extent to which its policy:

- is influenced by the views of Defense Department decision-makers and/or senior military officers
- flows from civilian officials' own presumptions that the military needs to carry exceptional weight
- assigns the military a leading role in implementing the nation's foreign policy, and
- treats military security and national security as if they were synonymous."

(Enloe, 2004, p. 122)

Consider the foreign policy of your own country. Who is making statements to the media about a particular issue: military officers, the Foreign Secretary (Secretary of State), or the Minister/Secretary of Defense? Does the Ministry of Defense/ Pentagon give regular news conferences?

assumptions about individual behavior that are then transferred to geopolitical codes. The essential ideological building block is the masculinity myth: the notion "to be a soldier means possibly to experience 'combat', and 'combat' is the ultimate test of a man's masculinity" (Enloe, 1983, p. 13; Hedges 2003).What it means to be a "man" and effective military operation are mutually reinforcing:

> Men are taught to have a stake in the military's essence – combat; it is supposedly a validation of their own male "essence". This is matched by the military's own institutional investment in being represented as society's bastion of male identity. That mutuality of interest between men and the military is a resource that few other institutions enjoy, even in a thoroughly patriarchal society.
>
> (Enloe, 1983, p. 15)

Combat defines the "man" and also validates the existence of the military. Moreover, combat as a masculine pursuit translates into the importance of the military as a masculine institution which, furthermore, plays a role in the militarization of geopolitical codes. The militarization of geopolitical codes is especially resonant when combat is in progress, defined as "likely," or a recent matter of national history. The foreign policy experience of "combat" defines the identities of individuals (men and women) and, hence, continues the relationship between the construction of individual identities and the form of geopolitical codes.

Combat, constructed as an essentially masculine pursuit, rests upon women in two ways. One is in the practical sense, the exclusion of women from combat duty but their necessary role of "camp followers" (Enloe, 1983): in other words, women play a number of "supporting roles" that are necessary for the military to function. Some of these roles are with the services, such as nursing and clerical work. Other roles are outside the services and even the law. Prostitution is perhaps the most obvious, but so is the role of the military or diplomatic wife (Enloe, 1990). Crucial to our connection of militarization, nationalism, and geopolitical codes are the twin needs of women's support services while women's roles are controlled and restricted to prevent "disorder," in the form of women's participation in combat. In the words of Cynthia Enloe:

> This mutuality of interests has the effect of double-locking the door for women. Women – because they are *women*, not because they are nurses or wives or clerical workers – cannot qualify for entrance into the inner sanctum, combat. Furthermore, to *allow* women entrance into the essential core of the military would throw into confusion *all* men's certainty about their male identity and thus about their claim to privilege in the social order.
>
> (Enloe, 1983, p. 15)

Combat defines the man, but man also defines the combat. In other words, by making combat the definition of manliness, and making it a male preserve, military combat is the defining event of a patriarchal society and its members. "Women may serve the military, but they can never be permitted to *be* the military" (Enloe, 1983, p. 15 emphasis in original). The militarization of geopolitical codes is enhanced as it serves individual goals

– making society's boys into men – while also facilitating a dominant role for the military in the definition of a geopolitical code. In turn, those with "combat experience," by definition men, are also privileged in public affairs (Enloe, 2004), a process very evident in US politics.

The militarization of society complements and intensifies the gender roles that are defined by nationalist ideology. Furthermore, in patriarchal societies "combat" has an essential role in the fundamental identity or purpose of men. Not surprisingly, war is a key ingredient in national myths and interacts with the gendered understanding of public and private roles. Not only are men the "defenders" of the nation, but actual defense is necessary to make a man. With the militarization of society, in addition to the role of "combat" in defining male identity, and the dominance of men, and the military, in public affairs, it is hardly surprising that the necessary construction of *difference* by nationalist ideology is readily "upgraded" into "hatred" and "threat"; in other words, war.

In the next section we discuss how different national histories can be related to different geopolitical codes and hence the forms of conflict that particular nationalisms may generate.

A typology of nationalist myths and geopolitical codes

The geopolitical codes of states rest upon the maintenance of their security. On the whole, security is related to the territorial integrity of the state. In other words, geopolitical codes define ways in which the sovereignty of the state must be protected or the state's status and well-being enhanced. Perceived threat of attack upon the citizens of the country requires a geopolitical code attending to boundary defense. Enhancing the status and well-being of a state often requires identifying historic grievances that have denied a country its "rightful" access to a particular set of resources. Consequently, an aggressive geopolitical code may be written that requires the seizure of territory.

Whether the code tends to be more defensive or aggressive, the concepts of sovereignty and territory remain central to the ideology used to justify the geopolitical code. Three types of "historical-geographic understandings" that frame the specific justifications of particular countries have been identified (Murphy, 2005, p. 283):

1   The state is the historic homeland of a distinctive ethnocultural group.
2   The state is a distinctive physical-environmental unit.
3   The state is the modern incarnation of a long-standing political-territorial entity.

These categories are not deterministic. Just because two countries possess a historical-geographic ideology emphasizing, say, territorial integrity, does not mean that they are likely to be equally aggressive. The benefit of this classification is that it shows that the justification for geopolitical actions used by a government must be grounded in a national ideology that resonates with the population; it must "make sense."

For example, the continuing conflict between Turkey and Greece is focused upon islands off the west coast of Turkey as well as the divided island of Cyprus. An interpretation of geopolitics emphasizing material pursuits would point to the oil reserves under

Turkey's western continental shelf. But what about the justification for the conflict? The Greek government's response to the Turkish invasion of Cyprus in 1974 is replete with allusions to modern Greece's unbroken connection to the ancient Greek empire. In the words of the Greek Foreign Ministry:

> The name of Cyprus has always been associated with Greek mythology (mostly famously as the birthplace of the goddess Aphrodite) and history. The Greek Achaeans established themselves on Cyprus around 1400 BC. The island was an integral part of the Homeric world and, indeed, the word "Cyprus" was used by Homer himself. Ever since, Cyprus has gone through the same major historical phases as the rest of the Greek world.
>
> (Quoted in Murphy, 2005, p. 285)

The connection between Greek gods and an estimated 225,000 refugees may appear tenuous to a neutral and objective observer. The point is that going to war could be justified to the Greek public, and gain support, through the usage of this widely held belief in the national history of the country.

More specifically, we can use the historical-geographical understanding of a country's geopolitical situation to suggest broad relationships between national identity and the content of a geopolitical code, though not in a deterministic sense (Murphy 2005, p. 286):

1   An ethnic distribution that crosses state boundaries is most likely to be a source of interstate territorial conflict where the ethnic group in question is the focus of at least one state's regime of territorial legitimation.

2   A boundary arrangement is likely to be particularly unstable where it violates a well-established conception of a state's physical-environmental unity.

3   States with regimes of territorial legitimation grounded in a preexisting political-territorial formation are likely to have particularly difficult relations with neighboring states that occupy or claim areas that are viewed as core to the prior political-territorial formation.

4   States that are not in a position to ground regimes in any of the foregoing terms are less likely to have territorial conflicts with their neighbors unless there are strong economic or political motives for pressing a territorial claim and state leaders can point to some preexisting political arrangement or history of discovery and first use that arguably justifies the claim.

The first point is the politics of nationalism discussed at the beginning of this chapter. The second point is illustrated by one of the most puzzling contemporary geopolitical tensions: the dispute between NATO allies Spain and Britain regarding the tiny territory of Gibraltar. Once of strategic importance, given its location at the mouth of the Mediterranean Sea, Spain's desire to gain control of Gibraltar is explained by the historical understanding of the physical extent of Spain; a physical geography currently violated by Britain's possession of just two square miles of territory. The third point has already been exemplified in a discussion of the Turkish–Greek conflict over Cyprus. In addition,

China's numerous territorial claims in East and Southeast Asia that rest upon the geographic extent of the ancient Chinese empire are a contemporary example of the third point.

The final scenario of geopolitical conflicts illustrates an important point; many states must legitimate their geopolitical codes without recourse to a national understanding of political geographies of ethnicity, physical extent, or historical claims. The states of Sub-Saharan Africa are colonial constructs; they are recent creations with little basis in ethnic homogeneity or physical legacies. Hence, there has been little cause for border conflicts in this region of the world. Instead, the geopolitics has been a matter of which ethnic group is able to seize control of the state apparatus, and not the geographical extent of the state (Herbst, 2000). In contrast, the imposed borders of Latin America shifted over the course of Spanish colonialism, creating opportunities for disagreement over their "proper" course.

Geopolitical codes are not simply an objective or strategic calculation made by foreign policy elites; it is not a matter of "statecraft" that excludes the majority of the population. Everyone is implicated, to a certain degree, because geopolitical codes cannot be enacted unless the majority of the population is acquiescent, at least tacitly. To ensure that a geopolitical code resonates with its citizens, a country is careful to frame its actions within the established political geographic sentiment of the nation's history. We have identified some broad categories with which we can interpret the major theme of the national tradition being evoked.

National identity frames the geopolitical acts of states within its commonly understood history. Emphasis is placed upon the important role that context plays in determining how and what people believe: "Identifying with a territory simply elicits certain views on the world, albeit in a contingent way, given certain national challenges, historical facts, and ideals" (Dijkink, 1996, p. ix). In other words, "to live within a territory arouses particular but shared visions (narratives) of the meaning of one's place in the world and the global system" (1996, p. 1). People are socialized within different territorial settings; what they hear, how they make sense of the information they receive, and the possible responses are limited by geographically specific institutions (Agnew, 1987). Referring back to our discussion of place in Chapter 1, it is the uniqueness of a country's geopolitical location or strategic situation, coupled with the way history is interpreted through dominant institutions, which formulates the particular ingredients of national ideology. Even in the age of satellite communication technology, and "globalization," information is distilled and interpreted through local journalistic and government lenses (Dijkink, 1996, p. 3).

Visions of one's country and its position in relation to other countries are formed within particular national myths. These myths form the basis for geopolitical codes and the means to represent and interpret these goals so that they obtain popular support. Dijkink's term is *geopolitical vision*: "any idea concerning the relation between one's own and other places, involving feelings of (in)security or (dis)advantage (and/or) invoking ideas about a collective mission or foreign policy strategy" (1996, p. 11). National histories are replete with the memories of both the pain of historical suffering and humiliation and the pride of past glories.

It is the tension between how these "maps of pride and pain," as Dijkink calls them, are remembered and used to initiate and justify foreign policy that makes geopolitical visions the way national sentiments are translated into geopolitical codes. "[N]ational identity is continuously rewritten on the basis of external events: and foreign politics does not mechanically respond to real threats but to constructed dangers" (1996, p. 5). Strategic concerns about resources and economics, and ideological referents to national values, combine in geopolitical visions, a framing of the world that connects the individual's sense of identity to global geopolitics through the geopolitical code of their country. The content of national myths and the content of geopolitical codes are made within dynamic contexts of conflict. The connections between national myths and geopolitical codes identified by Murphy and Dijkink show that geopolitical conflicts must be understood by connecting the actions of one set of geopolitical agents (those who control the state) with another group of geopolitical agents, the population of those states.

Activity

Identify which of the "historical-geographic understandings" we described at the beginning of this section, p. 115, best fit the way the nationalism of your country is portrayed. Think of current and past conflicts that your country has been involved in. Do the reasons and justifications of these conflicts follow the expectations of the framework? Why?

Breaking down the binaries

The previous section provided a typology to see how the construction of national myths has been essential in representing geopolitical codes in a way that makes them believable or readily accepted. Such representation requires the construction of us/them and inside/outside categories. The goal is to create a secure or stable sense of what is meant by "us" and how others are very different (see the essays in Giles and Hyndman, 2004). In other words, the nation requires an understanding that it is tidily bounded both physically and socially. The geographic extent of the nation is understood to be clear, it simply follows the lines on the map, and we are led to an understanding of who "belongs" or is a member of the nation and who is a foreigner, alien, or whatever term is used to describe an "other."

The dominant representation of the geopolitical world is one of clear distinctions, but the real world is quite different. Furthermore, alternative representations of the world have been advanced, especially by feminist geopoliticians (Giles and Hyndman, 2004), to encourage us to think in terms of loyalty and action outside the constraints of the nation-state framework. As we discussed at the beginning of this chapter, the term "nation-state" is a misnomer. Nation-states are not neatly bounded homogenous entities. Instead they are complex mixtures of different identities, which often spill over international boundaries. See the following case study of Burma/Myanmar as an illustration.

The important geopolitical implication of emphasizing the diversity of the nation and the way in which identities are not neatly compartmentalized within defined nation-state categories is that it requires us to rethink the notion of security (Giles and Hyndman,

2004). Defining and attempting to achieve security drives geopolitical practice and representation. The dominant framing of geopolitics through national identity means that security has come to be understood as national security, and threats are usually identified as the threats from other nations. In the current geopolitical context focusing upon terrorism the dominant rhetoric is still about national responses (we will talk about this more in Chapter 6). However, if we come to understand nation-states as false constructs and take that seriously then we must also question the idea of national security. It can be replaced with the idea of human security; that our commitment should be to the well-being of individuals regardless of their national identity, or religious beliefs, race, income level, gender, or sexuality. Not only does such an approach lead us to question the placement of the nation, or state, as the primary geopolitical actor, it also leads us to question how we should situate ourselves within collective identities and broader social groups; it forces us to give priority to other identities and groups than the state/nation. And by seeing security as something other than a national us/them competition we can consider a host of geopolitical actors that emphasize global connections rather than separation via national divisions.

As we will see in the following chapter on territory and borders, and discussions of networks (Chapter 6) and environmental geopolitics (Chapter 8), new geographies that emphasize flows between places and countries have become increasingly important in how we live and how we understand the world. The term "globalization" has come to be used as a catch-all phrase to describe the intensification of global interconnections and flows, whether it be money, commodities, people, or ideas. In Castells' (1996) terminology, we now live in a "network society." Unsurprisingly some scholars and commentators have predicted the "end of the nation-state." Such demise is proclaimed for two reasons. First, states can no longer manage or control global flows. In other words, they have lost their ability to claim sovereignty over a piece of territory. Second, people will become increasingly aware of global ties and develop a sense of identity that will transcend the nation; a global sense of collective identity will be established.

There is much skepticism that may be leveled at the "end of the nation-state" thesis. Especially, states are still powerful geopolitical actors, they can control flows across their borders to some extent, some global flows have been encouraged and assisted by states (especially financial ones), and the sovereignty of states has always been only partial. Also, the persistence of racialized and Oriental antagonism across the world is plain to see in many cases, and it is often framed through national perspectives. The vast majority of the world's population is not made up of cosmopolitan global travelers who are willing and able to be cultural chameleons. Most people live their lives within the same geographical setting and learn to see the world in that way. Consumer culture may sell global products, but the nation still plays a strong role in interpreting how we consume it. Global culture is adapted and understood within national settings. People are still readily mobilized to fight and die in the name of national security, and environmental disasters and wars may provoke global sympathy but the refugees they produce are still very much evaluated through an us/them and inside/outside national lens (Hyndman, 2003).

Though it is clear that the nation still plays an essential role in the practice and representation of geopolitics, it is also a mistake to view the operation of politics as being limited by discrete and clear-cut national boundaries. This mistake was labeled by Agnew

(1994) as the "territorial trap" or the dominant tendency to see processes of politics and society to be neatly encompassed within state borders. Instead, Agnew notes that sovereignty is "unbundled" through the operation of networks that cut across national boundaries. Networks of migration are a clear example. In the past few years the ability of global currency markets to dictate the way European countries manage their economies (Greece, Italy, and Portugal, for example) has illustrated how flows across the globe have constrained the sovereignty of states.

In sum, states operate in a world of global flows; and nation-states are a complex mixture of competing identities. Hence, the world political map of neatly bounded nation-states is a fiction. But it is an important fiction as it has been the material and representational basis of identifying "national security" imperatives that have been the driving force of geopolitics. By problematizing the nation-state to see it as a mixture of competing identities and by noting global connections through networks, dominant ideas of security can be destabilized to see the importance of human security within a connected global humanity (Hyndman, 2003).

Case study: Myanmar/Burma: a militarized state trying to build a unitary nation

This short case study of Myanmar/Burma is a précis of an excellent essay by geographers Carl Grundy-Warr and Karin Dean (2011) that explores how a militarized state has used force to create a sense of national unity in the face of ethnic diversity and opposition. The country was known as Burma until 1989, when the current regime changed the name to Myanmar. Both terms have existed in the history of the territory. Though the name change has been recognized internationally, the Burmese democracy movement has rejected it because they see it as part of the militarization strategy discussed in this case study. Myanmar/Burma gained independence from British colonial rule in 1948 and since then the national project has been one of a series of military–state actions to create a centralized state that reaches across the geographic expanse of the country. In March 1962 a military coup established the regime that continues to rule and deny democracy. State-building by the ruling military regime has been associated with fighting enemies, whether "internal" or "external." The Burma Army, or *tatmadaw*, has been the key institution in the attempt to create a clear sense of the "nation" (Selth, 1996). However, this national project has been continually challenged, even before the military coup, by ethnic and communist insurgencies in the border regions. In short, ever since the 1960s, and continuing today, the military has represented itself as the sole institution of law and order across the territorial extent of the state with the aim of creating a coherent nation-state and preventing the fragmentation of the country: the political goal has been preservation of "the Union" (Lintner, 1990; Grundy-Warr and Dean, 2011). In other words, the militarized state has attempted to create a sense of nation-state in the light of diversity and competing identities.

Since the coup, General Ne Win and the "Revolutionary Council" have established the Burmese Socialist Program Party (BSPP) as its party and created a militarized and isolationist version of the "Burmese Way to Socialism" that has attracted criticism from the international community for its human and civil rights violations. A key element of the government's project was the creation of a sense of "nationhood." One step was a political map which demarcated Burma as consisting of seven divisions surrounded by seven so-called "minority states." However, the map imposed by the government did not reflect how the "ethnic" political parties in the "minority states" identified the most appropriate form of federalist representation (Steinberg, 1984; Grundy-Warr and Dean, 2011). The BSPP's attempts to create a nation-state through force were challenged. Throughout the 1970s and 1980s the military regime and *tatmadaw* were in conflict with the Communist Party of Burma (CPB) and with a number of ethnic insurgencies that challenged the government's unitary vision of the nation-state.

Following the brutal repression of the pro-democracy movement in 1988 the State Law and Order Restoration Council (SLORC) was formed, a further militarization of the state with the continued goal of "national unity." The SLORC tried to establish political control through Law and Order Restoration Councils (LORCs) at different scales of administration, including state/divisional, district, township, and ward/village-authorities. With an eye to creating a unified nation-state the LORCs were planned to encompass all seven divisions (*taing* in Burmese) and the seven designated ethnic states (*pyi-neh* in Burmese) of the country previously created by the BSPP. In reality, the reach of the LORCs in the *pyi-neh* was patchy because of the ability of ethnic political parties and local armies outside the control of the military regime to establish rule and authority. In other words, geopolitical actors other than the state operated at local and regional scales to frustrate the military regime's attempts to create a unified nation-state.

After 1988 an attempt was made by the central government to accommodate ethnic groups through agreements and ceasefires. However, such attempts should be judged carefully and critically. Grundy-Warr and Dean (2011) argue that the agreements are just another means for the central government to extend its reach across the whole of the territory. Notably, it is argued that the purpose of these agreements has been to make it easier to exploit vital natural resources; especially oil, natural gas, teak, and gems. Given these goals, it is perhaps unsurprising that the establishment of a period of ceasefires has also seen the continuation of the military regime's brutal repression against pro-democracy uprisings in 1988 and 2008, especially the long-term suppression of the National League for Democracy (NLD) and Aung San Suu Kyi. The period of repression has continued to target "pockets of armed resistance, particularly in the west (targeting the Muslim Rohingya) and in the eastern borderland – targeting non-ceasefire groups, such as the Karenni National Progressive Party (KNPP), the Karen National Union (KNU), and remnants of the Shan State Army (SSA), particularly Shan State Army-South (SSA-S)" (Grundy-Warr and Dean, 2011, p. 94). These ethnic groups are a challenge to a unitary sense of the nation-state and it is for this reason that they have been targeted by the military regime and its national project.

As a result of the militarized national project the *tatmadaw* has seen a steady increase in terms of troop numbers and the geographic reach of operational deployments within the country (Selth, 1996; Grundy-Warr and Dean, 2011). Since 1990 it is believed that the army has doubled in strength to about 400,000 personnel, making it the twelfth largest military in the world (Selth, 2001: 12). Increasing numbers of troops are stationed in parts of the country where the military regime has established a presence either by force or by ceasefire negotiations. Sadly, the growth of the military has been accompanied by deterioration in public services, especially health and education, and a mismanagement of the economy – resulting in daily shortages and the expansion of corruption and black markets (Grundy-Warr and Dean, 2011). The military regime has ignored responsibilities to the daily well-being, or security, of individuals while ensuring a process of militarization – increased involvement in the political, economic, and social spheres. It is argued that these actions come from the military's geopolitical aim of military-centered state-building to achieve "National Unity" (Steinberg, 2007).

The geopolitical construction of a unified nation, or more accurately the attempt to create a nation-state in a diverse country, has included geopolitical practices such as the 2005 creation of a new capital, Naypyidaw, and subsequent annual military parades under the shadow of the enormous statues of the three historical Burman *unifying* kings (Grundy-Warr and Dean, 2011). The geopolitical representation of the SLORC's attempted construction of a militarized nation-state has required the broadcast of images of the "national geo-body" and the "nation." The key theme has been the identification of "Three National Causes": "the non-disintegration of the union, the non-disintegration of national solidarity, and the perpetuation of national sovereignty" (GOM, 1994). However, these are representational fictions hoping to justify state-building (Lambrecht, 2004).

The existence of ethnic groups within Burma, and their resistance to the central state, illustrates that a unitary vision of the nation-state is a fiction. This situation is common in the world, reflecting the artificial and imposed nature of state boundaries. The result is that governments try to create national unity by projects of state-building that sometimes include the construction of external threats. In the case of Myanmar/Burma the military regime has maintained a dynamic project of "national reinvention," in which the military-state and *tatmadaw* believe they must and can "hold the country together" and protect "national unity" (Callahan, 2004: 215–17; see Grundy-Warr and Dean, 2011). Part of this project has been the continuing suppression of the pro-democracy movement and, especially, Aung San Suu Kyi.

Summary and segue

Nationalism is an ideology which defines an overarching national identity that transcends ethnic differences. Simply put, the claim is that people within the boundaries of a state hold a common identity. In other words, the emphasis is upon homogeneity. Increasingly, however, emphasis is being placed upon hybridity. Individuals possess multiple collective identities, and, furthermore, these collective groups are themselves the product of mixture. The identity Arab-American, for example, is complicated by the complexity of both Arab and American, the way in which both reflect diverse experiences and identities. The same can be said for the term Black-British. Current discussion of the movement of people (legally and illegally, voluntary and forced) across the globe has, on the one hand, increased the hybridity of people's collective identity. On the other hand, some people have reacted to such movement by reinforcing a belief in maintaining the "purity" of national identity.

It is false to separate the "domestic" and the "foreign" in an understanding of geopolitics. The geopolitical actions of states, the way they interact with agents external to their boundaries, require the support, tacit or overt, of their populations. The ideology of nationalism provides a sense of loyalty to the state and the belief that security rests upon sovereignty and integrity of the territory to which a national group lays claim. A component of nationalist ideology is the promotion of gender roles that facilitate a militarized foreign policy. In this chapter we have seen the ideological "glue" that maintains states and their geopolitical codes. In the final sections we challenged the dominant view of the nation and state, and the false binary of inside/outside, to reconsider how we think of security. In the next chapter, we explore a geographic feature that is also essential in maintaining the integrity of states and their national identities: Boundaries.

Having read this chapter you will be able to:

- Understand the connection between national identity and the state
- Identify the manifestations of nationalism in current affairs
- Identify the way gender roles are defined by the practice of nationalism
- Identify the important role of "combat" in the creation of national identities
- Understand how geopolitical codes are rooted in national histories
- Question the dominant inside/outside binary underlying the geopolitics of national security

---

Further reading

Cox, K. R. (2002) *Political Geography: Territory, State, and Society*, Oxford: Blackwell.
   Parts of this textbook provide useful explorations of the state, including the local state.

Dahlman, C. (2005) "Geographies of Genocide and Ethnic Cleansing: The Lessons of Bosnia-Herzegovina," in C. Flint (ed.), *The Geography of War and Peace*, Oxford: Oxford University Press, pp. 174–97.
   This chapter explores questions that define a new field of inquiry, the geography of genocide.

Dijkink, G. (1996) *National Identity and Geopolitical Visions: Maps of Pride and Pain*, New York: Routledge.
   Makes connections between the national identity and foreign policy using a number of extended case studies.

Dowler, L. (2005) "Amazonian Landscapes: Gender, War, and Historical Repetition," in C. Flint (ed.), *The Geography of War and Peace*, Oxford: Oxford University Press, pp. 133–48.
   An example of Dowler's work examining the connections between nationalism, gender, and conflict.

Enloe, C. (1990) *Bananas, Beaches, and Bases: Making Feminist Sense of International Politics*, Berkeley: University of California Press.

—— (2004) *The Curious Feminist: Searching for Women in a New Age of Empire*, Berkeley, University of California Press.

Cynthia Enloe is the preeminent feminist scholar of militarism and militarization.

Murphy, A. B. (2005) "Territorial Ideology and Interstate Conflict: Comparative Considerations," in C. Flint (ed.), *The Geography of War and Peace*, Oxford: Oxford University Press, pp. 280–96.

The following readings provide further exploration of nationalism:

Anderson, B. (1991) *Imagined Communities*, second edition, London: Verso.

Billig, M. (1995) *Banal Nationalism*, London: Sage.

Smith, A. D. (1993) *National Identity*, Reno: University of Nevada Press.

—— (1995) *Nations and Nationalism in a Global Era*, Cambridge: Polity Press.

5

TERRITORIAL GEOPOLITICS: SHAKY FOUNDATIONS OF THE WORLD POLITICAL MAP?

In this chapter we will:

- Gain an understanding of the role boundaries play in geopolitics
- Define boundaries, borders, borderlands, and frontiers
- Situate boundaries and borders as one form of territoriality
- Consider territorial constructions other than states
- Discuss the role of boundaries in the construction of national identity
- Identify the boundary conflicts that most commonly appear in geopolitical codes
- Discuss how peaceful boundaries may be constructed
- Discuss the concept of the borderland and its implications for boundaries, nations, and states
- Provide case studies of the Israel–Palestine conflict and the Korean peninsula
- Introduce the territoriality of the sea

In the previous chapter we saw the geopolitical importance of the state; its representation as a nation-state, and the practices such representation requires and legitimates. An essential material and representational feature of states and nation-states is the boundary: the means to create in-groups and out-groups. The boundary is a material and ideological geopolitical feature. Despite eye-catching, or perhaps more accurately "book-selling," cries of the end of the nation-state and a borderless world, movement of goods and people (but less so ideas) is still constrained by physical controls imposed by governments. Much of the geographic work on the porosity of borders and boundaries has been by European geographers looking at the internal boundaries of the EU. Alternatively, the War on Terror has promoted fears of "porous borders," especially the US's own, plus its policing of the Afghanistan–Pakistan boundary and that of Iraq. Similarly, as the boundaries of the EU are relocated eastwards, the public pressure on European government to focus attention upon refugees and other immigrants increases. In sum, the geopolitics of borders and boundaries remains, but the geography is the product of strong imposition on the one

Box 5.1 Boundaries, algorithms, and your body

You are playing a role in the Global War on Terror just by buying a ticket, jumping on an airplane, and flying to a foreign destination. Your personal details and some of your actions are recorded in the process and used in a host of analyses that are used to create algorithms designed to identify terrorists as they attempt to cross international boundaries. The relevant social scientific term is "biopower" – or the way in which the personal details of individuals are used to create a host of legal and social categories: gender, race, religion, etc. These categories are then coupled with behaviors or decisions: what type of ticket was bought, when and by what means it was purchased, what meal and seat requests were made, etc. The belief is that authorities can use this information to identify individuals and the way they act in a politics of "risk management" that will prevent terrorism. The outcome is a grouping of categories that can be defined as, on the one hand, trusted traveler biometrics, and on the other hand, biometrics that can be used to identify people within categories of "threats" (Amoore, 2006, p. 343).

The issue for scholars such as Louise Amoore (a geographer at the University of Durham) is that "algorithms appear to make it possible to translate *probable* associations between people or objects into *actionable* security decisions" (2009, p. 52). Or in other words, the way that peaceable everyday travelers' actions (including yours) in the past are recorded and included in ever-updated algorithms that are deemed to make it possible to see into the future. As then US Secretary of Homeland Security Michael Chertoff (2006, p. A15, quoted in Amoore, 2009, p. 52) claimed:

> If we learned anything from September 11, 2001, it is that we need to be better at *connecting the dots* of terrorist-related information. After September 11, we used credit card and telephone records to identify those linked with the hijackers. But wouldn't it be better to identify such connections before a hijacker boards a plane?

But this is not a simple matter of science and statistics. Amoore (2009) points out that human judgment and labeling are essential to identifying individuals and groups as "risky." For example, surveillance cameras track "atypical" behavior that is judged and classified by a human, but then becomes translated into something that is seen as scientific and, therefore, less questionable. These algorithms are partially constructed through the insertion of technology into travel documents, such as passports and immigration forms. For example, radio frequency identification (RFID) is on trial at the US–Mexico boundary and can enable "smart" immigration and travel documents that may track the holder while in the US (Amoore, 2009).

Where is boundary control? One implication of the use of algorithms and RFID is the amorphous geographic and temporal location of the boundary. We contribute to the construction of boundary policing practices when we sit at home and buy a

(Continued)

ticket through our computer. We are being tracked before we even arrive at the airport. Visitors may be monitored as they go about their everyday lives miles and days after they have entered the country. As Bigo (2001) quoted in Amoore (2009) argues, such practices blur the distinctions between overt war (with its obvious violence) and war by other means that entails the control of the movement of people based on what they might, probabilistically, do based on attributes and behavior of others. As we shall see in this chapter, boundaries control people by creating in-groups and out-groups, or identities of us–them, the trusted and those to be feared. Boundaries are essential in this geopolitics of identity and control, but the "location" of these boundaries is something that is becoming increasingly vague and fluid.

hand, and greater porosity on the other. See Donnan and Wilson (1999) for an excellent discussion of boundaries and borders, as well as the collection of essays on specific boundary conflicts edited by Schofield *et al.* (2002).

In this chapter we focus on boundaries, but must note that their function is to control flow of movement. In the following chapter we shall concentrate on the geography of networks and the flows they facilitate. Boundary formation is one example of a broader set of geopolitical processes known as territoriality (Sack, 1986), or the way in which territories are used to impose political control. The territorial politics of the state and national idea are partially constructed through the geopolitics of boundary formation and control. Thinking of boundaries and states as just one form of geopolitical territoriality, and seeing this as a continually dynamic process, raises the question of how the state was formed (as discussed in the previous chapter) and other forms of territoriality; such as territories where no functioning state exists and supra-state territorial formations (e.g. the EU).

First, we will define our terms and examine the ways in which boundaries are created and maintained, and the geopolitical role they play. The broader geopolitics of territoriality will then be described. To help understand contemporary conflicts we will provide a brief catalog of potential border disputes by examining the woes facing the fictional country of Hypothetica. We will then exemplify our discussion with a case study of the Israel–Palestine conflict. Boundaries and borders are also geographical features that may reflect movements toward peace, the topic of the following section of the chapter. Finally, we will relate the demarcation of a boundary to global geopolitical conflict with a case study of the Korean peninsula.

Definitions

As with other topics, we will start by making sure we are using the same language, and we will adopt Prescott's (1987) terminology. The term *boundary* will be used to refer to the dividing line between political entities: The "line in the sand" if you wish that means you are in, say, Mexico if you stand on one side and the US if you hop over and stand on the other. Later we will look at the geopolitics of defining the precise location of the boundary and its effectiveness and role in controlling movement. The term "border" is often used synonymously with the term "boundary," but for our discussion it is useful to distinguish the two. *Border* refers to that region contiguous with the boundary, a region within which society and the landscape are altered by the presence of the boundary. When considering neighboring states, the two borders either side of the boundary can be viewed as one *borderland*. This is especially useful when looking at the cross-boundary interaction between two states.

Finally, a term that is often used in the media when talking about boundaries is *frontier*. To be precise, a frontier refers to the process of territorial expansion in what are deemed, usually falsely, as "empty" areas. For example, the American frontier involved the killing, expulsion, and confinement of Native Americans to facilitate the land's "settlement" and its integration into the US economy. Even when indigenous populations were recognized, the creation of a frontier was justified through the language of religion and civilization: the regional population was a void for Christian practices to fill and integrate into the Christian realm. Echoes of this language remain today, as failed states are identified as the repositories of "evil" and, hence, must be brought back into the international state system and its norms of behavior.

Modern geopolitics was the politics of boundary construction. The building block of geopolitics was the nation-state, a political geographic entity that required territorial specificity as the basis for its sovereignty. Boundaries delineated the population and resources that came under the control of particular states. The geopolitics of mapping modern boundaries has three stages (Glassner and Fahrer, 2004). First, the course of the boundary must be *established*. This decision can be made through war, mutual political

Figure 5.1 Closed border: Egypt–Israel.

Figure 5.2 Open border: Russian Caucasus.

agreement, or external imposition. For example, we will see the role of external states creating the boundary between North and South Korea in a case study later in this chapter. The political boundaries in the continent of Africa are overwhelmingly the result of decisions made by colonial powers (Herbst, 2000). Once the boundary has been established it must be *demarcated* – its course must be made visible. In some cases, the visibility may not be clear on the actual landscape, but is solely a feature of maps. One could walk across the boundary without knowing it. In some cases the visibility of the demarcation is sporadic; checkpoints exist at trans-boundary roads and railways, but a fence does not extend along the full extent of the boundary. In extreme cases, the demarcation of the boundary is a violent expression, a continuous barrier of concrete, razor-wire, land-mines, attack dogs, and trip-activated machine guns. Not surprisingly, the form of demarcation is related to the degree of *control*, the third and final component of mapping boundaries. Decisions about the nature and intensity of flows across a border display great variation. North Korea is the most "closed" of all the contemporary states; goods, people, and information rarely travel out, and the opposite flow is sparse and completely controlled by the government. In the United Kingdom, entrance from other EU countries is relatively free, but there are many restrictions, made as visible by the government as possible for political capital, of refugees. The degree of control also varies with time; post 9/11 travelers entering the US have come under much more rigorous inspection, and required documentation has increased.

With so much effort being put into the establishment, demarcation, and control of boundaries, one must reflect upon the geopolitical purposes that boundaries serve. Within the geopolitical context of the War on Terror, boundary control is related to "security." States maintain their legitimacy, in part, by keeping their citizens safe, and control of borders is a pivotal factor. For example in the US the Office of Homeland Security was established in the wake of the terrorist attacks of September 2001 in order to enhance boundary security. Another example is Israel's success in establishing its boundaries in its quest to provide a territorial haven for Jews in a policy of Zionism.

The connection between boundaries and security is more complex than the ability to prevent invasion or infiltration. National identity is a territorial identity that rests upon the existence of, or desire for, a state with sovereignty over a piece of territory. National homeland, mythologized as it is, and state authority both rest upon territorial demarcation; boundaries demarcate nations and states and so define nation-states. Boundaries are, simultaneously, instruments of state policy, the expression and means of government power, and markers of national identity (Anderson, 1996). Their role in providing security extends into the taken for granted nature of national identity and citizens' expectations of government services.

The converse becomes of interest in discussions of the porosity of boundaries. If boundary control is, at least in some regions of the world, increasingly beyond the control of states, then what are the implications for national identity and state authority? We will address this geopolitical development later in our discussion of borderlands.

Constructing territory

Our discussion of the process of boundary construction shows us that territory is constructed by geopolitical agents (Elden, 2009). In the case of the boundaries of colonial Africa, for example, or the demarcation of the ceasefire line between North and South Korea that became recognized as an international boundary, external powers defined the course of boundaries. Over time these established boundaries became recognized features, or structures, of new geopolitical activity. The broader point to consider here is that all forms of territorial politics are the product of agency. For example, the simple act of a homeowner erecting a fence to keep a neighbor's dog off the lawn is an act to declare a particular piece of territory "off limits."

Geographer Robert Sack (1986) has called such processes territorialization, or the way that territory is used to enable politics. One of the clearest and most dominant forms of territorialization in the study of geopolitics is the processes of state formation that we discussed in the previous chapter. Jean Gottman (1973) contrasted two expressions of geopolitics that are useful in understanding the importance of territory: flows between spaces and bounding space into identifiable areas that structure how we live. Some forms of geopolitics are best thought of as movement of "things" across the globe. We live in a world where we expect and assume that the products we buy in shops will likely have come from across the globe; students are encouraged to "study abroad" and frequently share classrooms at their "home" institutions with students from foreign countries; investment firms shift money in pension plans across the globe, using the temporal sequencing of financial markets across the world's time zones to create a constant flow; immigrants and refugees are a focus of vociferous political arguments; and finally terrorists have been identified as a threatening flow that requires a securitization of boundaries.

The geopolitics lies in the discussions of how, and to what extent, the amount and speed of these flows can be controlled. This form of geopolitics is the second part of Gottman's idea and has centered upon the ability of states to create a territorial politics that not only controls flows but creates a view of politics that is bounded or limited by loyalty to the nation-state. The latter is the politics of nationalism that we discussed in the previous chapter. The former is the never-ending tension between the desire for state boundaries to open to some degree and for some purposes but closed for others. Those who want strict restrictions on in-coming migrants or refugees would still like to leave their country and be allowed to enter another on vacation. Though some may want restrictions on some products entering their country, such as rice or other agricultural goods produced more cheaply abroad, they are also used to cars and electronic products being relatively accessible because of free-flowing global trade. The different interests and opinions toward restricting or allowing different forms of flows mean that the geopolitics of the territorial restrictions of flows is dynamic and often contradictory.

In the next chapter we will concentrate upon the geopolitics of flows. In this chapter our emphasis upon boundary formation and management as a territorial process suggests that there are other geopolitics of territorialization (Elden, 2009). We may consider two questions about the contemporary geopolitics of territorialization. Has the geopolitical project to cover the whole of the globe with territorialized nation-states regressed? What alternative forms of territorialization are emerging?

The first question forces us to consider the geopolitics of deterritorialization, or the way in which what was believed to be a coherent territorial nation-state loses its ability to enact the despotic and infrastructural forms of power we introduced in the previous chapter. In contemporary geopolitics these entities have been represented as "failed states" and have been identified as security threats (Patrick, 2007; Clunan and Trinkunas, 2010). The definition of a failed state is contested but revolves around the inability of a central government to rule effectively across the whole of its territorial extent. Not only is a state unable to provide basic services (especially education and health); it has no ability to provide order or security for its population. Instead, geopolitical actors that are represented under various labels (such as rebels, warlords, and terrorists) display effective rule in different parts of the country. These states – Somalia and Yemen are often cited as prime examples – have been identified as geopolitical threats within the United States' geopolitical code, represented as the Global War on Terror. As we will discuss in Chapter 6, the United States has identified "failed states" as potential "safe havens" for terrorists, especially al-Qaeda. Immediately after the terrorist attacks of September 11, 2001 Afghanistan was represented as a state in which Taliban warlords held sway and facilitated the presence of senior al-Qaeda leadership.

In contrast to the notion of deterritorialization, or framing politics within state boundaries as somehow "failed," there are ongoing processes of reterritorialization. The term reterritorialization is used to consider how territorial geopolitical entities other than nation-states are becoming increasingly important. The most obvious example is the EU, a territory made up of an organized grouping of states. Other parts of the world have also

Box 5.2 Connecting failed states and human security

In 2011 the World Bank issued a report outlining its concern about the inability of some states to provide for their citizens (World Bank, 2011). The report estimates that 1.5 billion people on the planet live within situations of inadequate state rule and continually experience violence and criminal activity. In the words of the report: "How is it that almost a decade after renewed international engagement with Afghanistan the prospects of peace seem distant? How is it that entire urban communities can be terrorised by drug traffickers? How is it that countries in the Middle East and north Africa could face explosions of popular grievances despite, in some cases, sustained high growth and improvement in social indicators?"

These are unsettling questions and are a stark illustration of the differences in life experiences between the relatively comfortable and the vulnerable in today's world. They are also examples of the ways in which some states are failing to provide a context for basic security in daily life. The report also highlights the contemporary security agenda that downplays conflicts between states and emphasizes civil wars, as well as more fluid and less easily defined civil disorder. The intersection of crime and politics is also a concern. The report highlights the case of Guatemala, where the levels of violence related to crime and drugs have surpassed the killing during the country's civil war of 1980–95.

The focus of the World Bank has always been economic development rather than conflict. Hence the report makes a connection between security and economics, arguing that the lack of employment prospects in many countries lies at the root of instability. The World Bank suggests that any steps to improvement are likely to be gradual: a generation of institution building that will lead to the three critical ingredients of security, justice, and jobs is seen as the way forward.

The positive aspect of this report is the emphasis upon human security that we introduced in the previous chapter through the framework of feminist geopolitics. The more pessimistic can point to the persistence of differences between global North and South and suggest that continued focus upon the individual scale will remain necessary if human, national, and global security is to be achieved.

The report can be downloaded at http://wdr2011.worldbank.org/sites/default/files/WDR%202011Overview_0.pdf.

seen tendencies for states to come together and cooperate – the African Union and the Association of Southeast Asian Nations (ASEAN) for example. The EU is the most important example, though, because the member states have created laws and institutions that have territorial reach beyond their own boundaries to encompass the whole Union. Employment laws, regulations about business monopolies, human rights, and environmental laws all have a territorial expression across the whole of the Union, and national laws must be altered to reflect the supremacy of the laws and regulations of the EU. The freedom of movement of citizens of EU countries throughout the whole of the Union, as well as the establishment of a common currency (the Euro), are the best examples of this reterritorialization. However, the resistance of Great Britain, an EU member, to the Euro – it has retained its own currency – and the continued existence of some passport checks at state boundaries illustrate that the reterritorialization of European politics is contested and member states still enact their abilities to restrict flows across their boundaries. Furthermore, in contrast to allowing freedom of movement within the territorial extent of the EU, there is a related territorial politics of restricting migration from other countries, especially from nearby Africa.

Geopolitical codes and boundary conflicts

Though we are currently witnessing the intertwined and dynamic processes of deterritorialization and reterritorialization, boundary geopolitics is still an important issue. Boundary conflicts remain a key motivation for states to go to war or make threats to do so. Figure 5.3 shows the sorry situation of a fictional country Hypothetica, a country that suffers from most of the usual grievances over boundary issues that can ignite conflict (Haggett, 1979). The separate issues can be grouped into four main categories: identity; control of national resources; uncertainty over demarcation; and security.

Identity

In discussing our definitions and functions of boundaries, we saw that they play an important role in the geopolitics of nationalism. Nations require or desire the establishment of boundaries; they provide the legitimacy and power of the state. The geopolitics of an internal separatist movement reflects a perception that a group within Hypothetica has identified itself as a nation separate and different from Hypotheticans. For the separatists, the boundaries of Hypothetica do not provide a meaningful territorial marker for their national identity, and the boundary needs to be redrawn so that a new nation-state is created. The geopolitics of such a boundary dispute are likely to be difficult to resolve and the potential for violence is high, because the separatists' attempt to define national boundaries is an attack upon the notion of territorial integrity of Hypothetica, an integrity that is the basis for its state power and national identity. The geography of the dispute also heightens the difficulties. The location of the separatists wholly within Hypothetica disrupts two related understandings of nation-states: a common nationality within the state's boundaries, and the territorial integrity of the nation-state.

The same issues exist for Hypothetica in two other locations. An ethnic group, with a collective identity distinct from both Hypotheticans and their neighbors, straddles the boundary. The primary collective allegiance of the ethnic group is not Hypothetican or the national identity of its neighbor. Perhaps, the establishment and demarcation of the boundary ignored the location of this ethnic group, or decided that it was insignificant. On the other hand, the ethnic group may only have mobilized its identity into a political issue once the boundary had been established, and the control of the boundary prevented interaction between members of the ethnic group, patterns of interaction that were likely to have been established in the group's culture.

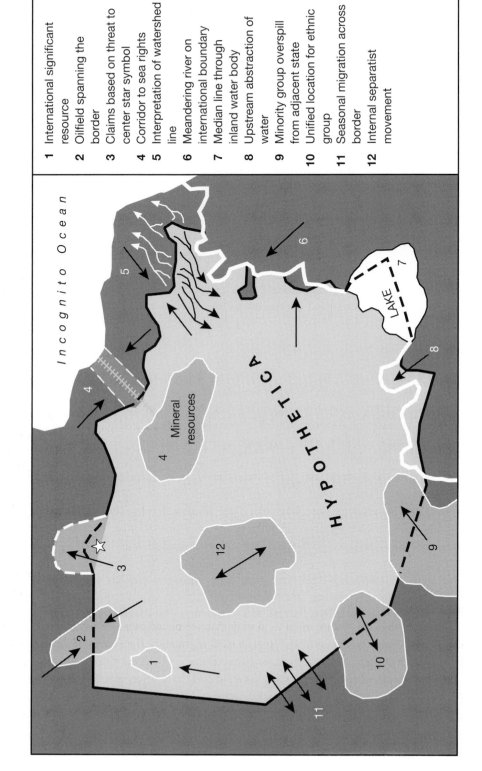

Figure 5.3 Hypothetica.

Legend:

1 International significant resource
2 Oilfield spanning the border
3 Claims based on threat to center star symbol
4 Corridor to sea rights
5 Interpretation of watershed line
6 Meandering river on international boundary
7 Median line through inland water body
8 Upstream abstraction of water
9 Minority group overspill from adjacent state
10 Unified location for ethnic group
11 Seasonal migration across border
12 Internal separatist movement

A similar problem exists to the northwest. The imposed boundary of Hypothetica transects historically established patterns of seasonal migration of pastoral peoples, following a path determined by the changing seasons and physical landscape in the search for water and fodder for their herds. The boundary does not take into consideration the functional needs of the pastoral peoples; their seasonal movement (or flow), possibly seen as "primitive," runs counter to the modern definition of nation-state spaces. In some instances, states may be unable to control such flows, or deem the seasonal movement as unimportant to national security. In other cases, the control of the movement may heighten as the geopolitical context changes, disrupting the social geography of the pastoral group.

The final boundary issue related to identity facing Hypothetica is a matter of the boundary's imprecise reflection of the geography of national identity. A minority group within Hypothetica has been created – a group that identifies with the national identity of the neighboring state. Political campaigns to unite such groups with the neighboring national body are known as irredentism. As we saw in the discussion of nations and states, such situations may result in pressures by Hypothetica to expel the minority group and/or attempts by the neighboring state to redraw the boundary and capture some of Hypothetica's territory so that the minority is no longer outside the boundaries of "its" nation-state.

Demarcation

Demarcation of a boundary often reflects the physical geography of the landscape. Indeed, as we discussed in the typology of national myths in the previous chapter, physical coherence may be the ideological basis of the nation-state. The physical barrier imposed by mountain ranges has led them to be used as the basis for political boundaries, but this can result in an imprecise and disputed boundary demarcation. Logically, if a mountain range is to act as a boundary then the "center" of the range should be pinpointed. The physical center of the range is the watershed line, the line that divides the process of precipitation run-off; in other words, if a raindrop falls on one side of this physical line it will flow, say, east, but if it landed the other side it would flow west. In theory, this physical feature is definite and precise. In practice, especially in remote and rugged terrain, it is hard to define and demarcate across the whole extent of the mountain range or political boundary. Uncertainty in the course of the watershed line can result in different interpretations of the course of the boundary, resulting in conflicts regarding demarcation.

Another physical feature often used to demarcate boundaries is a river, and often the thalweg or deepest channel of the river is used to pinpoint the course of the boundary. However, rivers are highly dynamic physical features. The flow of the water through the landscape creates erosion of the banks, and the course of the river will change over time. If the river has been used to demarcate a boundary, does the political boundary follow the old or new course of the river? If the old course of the river remains the official line of the boundary, what practical problems regarding fishing, agriculture, and water rights, for example, will emerge?

The final issue relating to physical features and boundary demarcation involves the use of lakes. If the boundary between states cuts through a lake, the norm is to define the median line between the shores as the boundary's line. However, erosion and changing water levels can provoke conflicts over the line's course, and the inability to paint a line on the water can lead to problems of control; precisely where does one state's jurisdiction end and the other's begin?

Resources

Boundaries define the territorial extent of a state's sovereignty, and sovereignty includes the right to extract and use resources. The course of a political boundary decides which states have access to which resources, and which states do not. Three resource-related boundary issues are facing the sad and troubled Hypothetica. First, on the southern border, water resources are a concern. The neighboring state is upstream, meaning the land in that state is higher in altitude and the water travels through it before reaching Hypothetica. The water in the river is available for use and misuse before it crosses the boundary and reaches Hypothetica. The upstream state could, for example, use all the water in the river for irrigation or industry leaving the river dry and denying Hypothetica use of the water. Also, the upstream state could pollute the river, not only denying Hypothetica use of the resource but delivering it a problem of toxic waste and environmental risks.

In the northwest of Hypothetica an oilfield spans the boundary. Who has access to the oil, and, more specifically, how should the quantity of oil in the reserve be divided between the states? Next to the oilfield is a deposit of a particularly significant resource, uranium. Given the importance of this resource to the rest of the world, Hypothetica may face pressures to extract and sell it in a particular way. For example, uranium, essential for making nuclear weapons, is a resource that lies beyond the control of the state in which it is located. International agreements control how much, to whom, and for what purpose, reducing the effective sovereignty the country has over its ability to sell uranium.

Security

The final set of boundary issues facing Hypothetica, a country I strongly recommend you do not invest your life savings in, falls under a general title of security. Hypothetica is a land-locked state, and so depends upon the goodwill of its neighbors to import and export goods by land. Particularly, the transport of mineral resources requires access to the sea,

---

Activity

Look through an atlas of contemporary conflicts, such as Andrew Boyd's *Atlas of World Affairs*, and see if you can relate the boundary conflicts identified in Hypothetica to real-world conflicts. In what way do the different types of boundary conflicts interact? Also, by looking at one conflict in detail, think about how different social groups (class, race, gender, state bureaucrats, etc.) have different roles in these conflicts.

and so Hypothetica may negotiate for a territorial corridor to the ocean. Conflict can result if the corridor is not granted, is controlled, in the eyes of Hypothetica, too rigorosly, or is closed once established. Finally, in light of potential or actual conflict with its northern neighbor, Hypothetica has invaded and now controls some of the land of its northern neighbour: the justification being that rocket or guerilla attacks on town "A" were emanating from across the boundary, as in the case of Israel and its boundary with Lebanon and the Gaza Strip.

Case study: Israel–Palestine

The boundary of the state of Israel is often in the news, especially with regard to the Palestinian Authority, though low-level conflict continues with Syria and Lebanon too. Perhaps more than any other contemporary geopolitical issue, the conflict between Israel and the Palestinians is fought with "facts" as well as tanks and thrown stones. The history of the dispute is contested by each side in order to portray their current actions as just. Here, I will try and give a "bare bones" history of the dispute in order to help us understand contemporary developments. I am sure it will not be to the satisfaction of anybody deeply committed to either side, but that is not its goal. I merely hope to provide some background to allow a reader who does not have a deep knowledge of the conflict to interpret media reports and also begin their own exploration of its causes, claims, and counter-claims. For a more in-depth discussion there are numerous sources, and each one will be perceived as biased. Well, they are. Here is my pick: Shlaim's *The Iron Wall* and *War and Peace in the Middle East*, Friedman's *From Beirut to Jerusalem*, Bregman and El-Tahri's *Israel and the Arabs*, Drysdale and Blake's *The Middle East and North Africa*, and Mansfield's *The Arabs*. Also, to weigh the opposing views, compare Said's *From Oslo to Iraq and the Roadmap* with Netanyahu's *A Durable Peace*. The following history is my attempt to use these sources, and some additions, to highlight the historical basis of the contemporary conflict.

Similar to many boundary conflicts and related nationalist struggles, the Israel–Palestine conflict began with the dissolution of an empire. The Ottoman Empire was first established in the mid-1400s and at its peak had extended into Europe, across southwest Asia, and parts of north Africa. However, by the end of the nineteenth century it was in terminal decline. On the one hand, this resulted in some of its subjects seeking greater autonomy and independence. On the other hand, powerful countries such as France and Great Britain were extending their influence into Ottoman territory, as we saw in Chapter 3 with regard to Great Britain and Kuwait. The decline of Ottoman power was provoking both internal and external interest in establishing boundaries in territory that had been or still was under the declining control of the Empire.

At the same time, the ideology of Zionism was gaining momentum. Zionism was creating a sense of Jewish national identity. It was a secular nationalism, with elements of socialist ideology, and its tenets were captured in Theodor Herzl's *The Jewish State* (1896). From our earlier discussion of nationalism we know that as in all nationalist movements a necessary connection between nation, state, and territory is made. Though other parts of the world were floated as possible sites for a Jewish state, the main focus was upon the biblical lands of Israel. The convening of the First Zionist Congress in 1897 encouraged and promoted Jewish migration to Palestine in a policy that was defined as "a people without a land for a land without people." This statement is the kernel of the current conflict, for at the time more than 400,000 Palestinian Arabs lived in Palestine. However, within thirty years, Jewish immigrants outnumbered the Palestinian Arabs.

In World War I, the Ottoman Empire was one of the axis powers and the region was a key strategic theater. Though the allies had defeat of the axis powers in mind, there was also considerable rivalry and scheming on the side of the allies. France and Britain used the war to jockey for position in a struggle between the two of them for greater control in the region after the war. There was much duplicity and tension, between the French and the British, and between the two European powers and the Arabs with whom they tried to foster alliances. During the war, in 1917, British Foreign Secretary Arthur Balfour, in what became known as the Balfour Declaration, stated:

> His Majesty's Government view with favour the establishment in Palestine of a national home for the Jewish people, and will use their best endeavours to facilitate the achievement of this object it being clearly understood that nothing shall be done which may prejudice the civil and religious rights of existing non-Jewish communities in Palestine, or the rights and political status enjoyed by Jews in any other country.

The two halves of the statement contradicted each other, as there was no plan on how a Jewish state could be established without compromising the existing Arab residents. In the wake of the Balfour Declaration and continued Jewish immigration, Arab–Jewish violence began around 1919. In one incident in 1929, fifty-nine Jews were killed in Hebron. The volume of Jewish immigration increased in conjunction with lobbying efforts by Zionist organizations in France, Britain, and the US. In what was interpreted as a pro-Zionist move, the British government appointed a Jew and Zionist, Herbert Samuel, as governor of Palestine; a territory it now controlled in the wake of World War I.

Partly as a result of these developments, an Arab rebellion lasted from 1936 to 1939 in which 5,000 Arabs were killed, some as a result of aerial bombing by the Royal Air Force. The initiation of World War II altered Britain's geopolitical calculations. Britain needed cooperation from the new Arab states and territories to secure the flow of oil, and, more importantly at this time, to maintain a continuous territorial link with British India. As a way of nurturing Arab support, the British decided to try and reduce the flow of Jewish immigration to Palestine to a trickle.

Unsurprisingly, Britain's policy met with resistance from the Jewish migrants. The policy was especially hard to justify in light of the Holocaust, Hitler's persecution of the Jews. At the end of World War II British policy was in tatters. Promises had been made to wartime Arab allies to limit Jewish immigration, but the Holocaust had energized the moral argument of Jews for a state. The Zionist movement resorted to violence, defined as terrorism, resistance, or national liberation, depending upon the political vantage point. A turning point was the bombing of King David Hotel in Jerusalem in 1946, the headquarters of the British administration, in which ninety-one people died.

The intensity of the campaign led Sir Alan Cunningham, senior British official in Palestine, to admit the "inability of the army to protect even themselves." The 100,000 British soldiers in Palestine were unable to control the 600,000 Jews living there. The threat of violence toward the soldiers was so great that the troops were ordered to stay within their compounds, unless they left in groups of four with an armed escort. In a defining moment, two British sergeants left their compound, were captured by terrorists of the Irgun group, and were killed, with their bodies displayed for public viewing. The British government and people had had enough and handed over the situation to the UN.

The UN drew up a partition plan in November 1947. Under the plan, a Jewish state would control 56 percent of the existing Palestinian mandate, and an Arab state would control 43 percent. The city of Jerusalem would be a UN-administered, internationalized zone. The plan left no one happy. The Zionists were upset as the Jewish state would not cover the whole of Palestine, as per the Balfour Declaration. Arabs saw a grave injustice, with Israel receiving 56 percent of the territory, when Jews accounted for just one-third of the total population and owned just 7 percent of the land. Despite some misgivings, the Zionists accepted the partition plan, which was a generous territorial award and led to the recognition of the state of Israel.

In 1948, in the wake of the British withdrawal and the presentation of the partition plan, war broke out as the contiguous Arab states, plus Iraq, invaded to negate the establishment of the state of Israel. The war was a victory for the Israelis: they ended up with all of the area allotted to a Jewish state by the UN plan, plus half of that allotted to the Arab state. Jordanian forces, with the help of Iraq, held the "West Bank" of the Jordan river and Egyptian forces held the "Gaza strip." About 700,000 Arabs became refugees in Gaza, Jordan, and Lebanon, and approximately 700,000 Jewish immigrants arrived in Israel over the following twelve months. Simply put, the war of 1948 created the "de facto" boundaries of the state of Israel. East Jerusalem was controlled by Jordanian forces, but Israel proclaimed it as its capital, a move that was not recognized internationally.

In the decades that followed, a series of wars demarcated and established Israel's boundary with its Arab neighbors. In 1956, Egypt "nationalized" the Suez Canal, sparking an invasion by British, French, and Israeli troops who feared that Arab control of the canal would disrupt use by Western countries of this vital trade route. The occupation was short-lived, however, as the troops were forced to withdraw under intense US displeasure.

Two other wars followed that were more of a success for Israel and established the continued military dominance of Israel in the region as well as Israel's current boundary. The Six Day War of 1967 began when Egypt moved its army up to the Israeli boundary and blockaded the Gulf of Aqaba. In response, Israel attacked Egypt, Jordan, and Syria and easily captured the West Bank, Gaza, the Sinai Peninsula, and the Golan Heights. After this violent re-demarcation of the Israeli boundary, the Arab states responded with what is known as the Yom Kippur war of 1973 when Egypt and Syria attacked Israel over the religious holiday. Despite initial successes given the element of surprise, the Syrian army was soon defeated and an Israeli counter-attack encircled the Egyptian army.

The Yom Kippur war was a turning point in the conflict, though not a decisive one. Defeat led the Arab countries to reconsider the benefits of the relationship they had established with the Soviet Union, with the region being a strategic focus of the Cold War. While the UN brokered a gradual Israeli withdrawal from the Sinai Peninsula, President Anwar Sadat of Egypt turned away from the Soviet Union and began to explore peace with Israel – a very brave initiative for any Arab leader. The result was the 1978 "Camp David" peace agreement that ushered in massive and continuing US aid to Egypt and Israel, but established the first peace agreement between Israel and a neighbor.

Of central significance to the conflict and hopes of a resolution are the UN resolutions 242 and 338 passed after the 1967 and 1973 wars respectively. The key points of the resolutions are:

* Israeli withdrawal from the occupied territories, meaning the West Bank and Gaza;
* recognition of the state of Israel and an end to the state of conflict;
* the right of return for Palestinian refugees being left vague and open to competing interpretations.

These resolutions have been the basis for the Palestinians' claims to the West Bank and the Gaza Strip; what they see as the necessary territorial foundation for a Palestinian state. With, at least in theory, goals of their own nation-states living peacefully side-by-side, the Israeli state and the Palestinian people have attempted peace negotiations. These negotiations have been sporadic. At times, expectations and hopes of peace have been high, but at other times the situation has been confrontational. Israel, an independent state with a large and sophisticated military, has dominated the Palestinians in terms of the ability to create "facts on the ground": a codeword for putting its military and people where they want to, despite their illegality under international law, diplomatic protest, Palestinian stone-throwing and civil disobedience in an *intifada* or uprising, and terrorist attacks upon Israeli citizens by factions of the Palestine Liberation Organization (PLO) and Hamas.

Attempts at peace have followed three general rubrics. "Land for Peace," or the "two-state" solution, calls for Israel to comply with resolutions 242 and 338 and withdraw from the occupied territories of West Bank, the Gaza Strip, and East Jerusalem. The withdrawal would, so the story goes, be the basis for Palestinian national self-determination and sovereignty. Though this is often portrayed as a huge success or victory for the Palestinian people, the historical timeframe we have adopted in this case study shows that such a move would be seen as an enormous compromise by the Palestinian people: they would gain control of just 23 percent of what they see as their historic homeland. Put the other way, Israel would control 77 percent of the land covered by the Palestinian mandate.

The belief exists that "Land for Peace" would lead to "comprehensive peace": in other words, once Palestinian people achieve national self-determination then the Arab states would recognize the state of Israel and make peace once and for all. Though countries such as Egypt and Jordan have made steps along this path, continuing conflict, at various levels, with Syria and Iran, for example, suggests that the connection or path should not be taken for granted. The harshest "plan" that exists is the Israeli rhetoric of "peace for

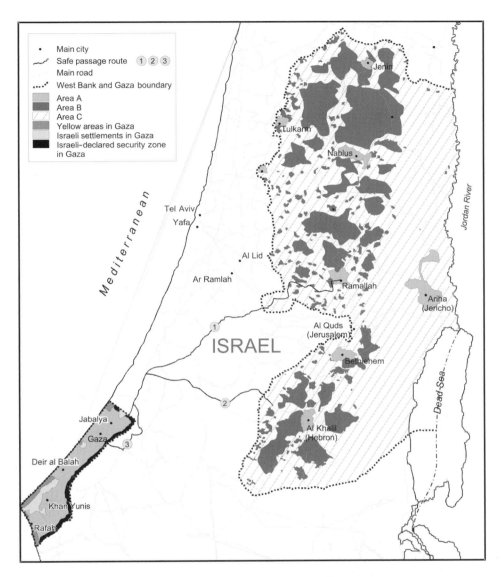

Figure 5.4 Israel–Palestine I: Oslo II Agreement.

peace", or a construction of the conflict as Israel's self-defense against an untrustworthy enemy that is not worthy of the title "negotiating partner." From the Palestinian perspective, such a stance not only is being cavalier with history but fails to acknowledge the level of violence committed against the Palestinians – deemed greatly disproportional to the Israeli's loss from terrorism (Falah, 2005).

"Land for Peace" rests upon Palestinian control of the West Bank. But what does "control" of the West Bank mean? Negotiations have given some concrete basis to who would control what in the West Bank, though the points of negotiation are contested within both the Israeli and the Palestinian camps. Under what are known as the Oslo II Agreements of 1995, the West Bank has been divided into three areas:

- Area A: controlled by the Palestinian Authority;
- Area B: Palestinian civil control and Israeli security control;
- Area C: Israeli authority.

Under closer scrutiny, such a division strongly favors Israel. Area A comprises just 3 percent of the West Bank, Area B 27 percent, and Area C an overwhelming 70 percent. In addition, Israel would maintain control of East Jerusalem. The geography of the division results in a "Swiss cheese" state for the Palestinians, the small area they control being surrounded by territory under the control of the Israeli military (Falah, 2005). It would be like controlling Cardiff and Swansea, or Madison and Milwaukee, but not being able to move freely between them. The balance of power, or the element of territorial control, would remain firmly in the interests of Israel as they cite terrorist threats and overarching hostility to the state of Israel. As Prime Minister Sharon said, on January 28, 2003: "Palestine would be totally demilitarized . . . ; Israel will control all the entrances and exits and the air space above the state; Palestinians would be absolutely forbidden to form alliances with enemies of Israel."

The death of Yassar Arafat, long-term leader of the Palestine Liberation Organization and first chairman of the Palestinian Authority, was believed to offer opportunity for progress toward peace. The election of Mahmoud Abbas in January 2005 as new President of the Palestinian Authority was met with hopes for peace, but also brinkmanship from Prime Minister Sharon, who threatened renewed occupation of Gaza unless terrorist attacks were halted. Abbas must satisfy both the Israelis and Palestinian militants; a tough task that will require the US and other influential countries to encourage talks and ensure that both sides act in good faith.

However, while attention is often drawn to terrorist attacks by Palestinians, the Israeli government is using its dominant position to alter the geography of settlement and occupation that will (i) make Abbas' task of bringing militants into the political process very difficult, and may increasingly alienate the mainstream, and (ii) create "facts on the ground" that run counter to the spirit and goals of the UN resolutions, and (iii) represent violations of international law as well as of human rights.

Specifically, though attention has focused on Sharon's promise to remove Israeli settlers from Gaza (a promise fulfilled in August 2005), the process of establishing settlements in the more desirable and historically significant West Bank continues. International law prevents countries from building permanent structures and communities on land that they control through military occupation. Since the Oslo peace accords of 1993 and 1995 the amount of settlers on the West Bank has more than doubled to 200,000, including thirty new settlements. Of course, the different sides of the conflict will portray this settlement in different ways. On the one hand, the claim is made that Jewish settlements constitute only 1.7 percent of the land of the West Bank. However, when the full extent of the municipal boundaries is considered, as well as the territorial extent of the authority of Jewish regional councils, then the coverage extends to 6.8 percent and 35.1 percent respectively (Falah, 2005).

The increase in settlements is paralleled, and some would say facilitated, by the building of "The Wall" or Security Fence along a route that is based upon the "Green Line" boundary, but with some key exceptions (Newman, 2005). The Israeli government

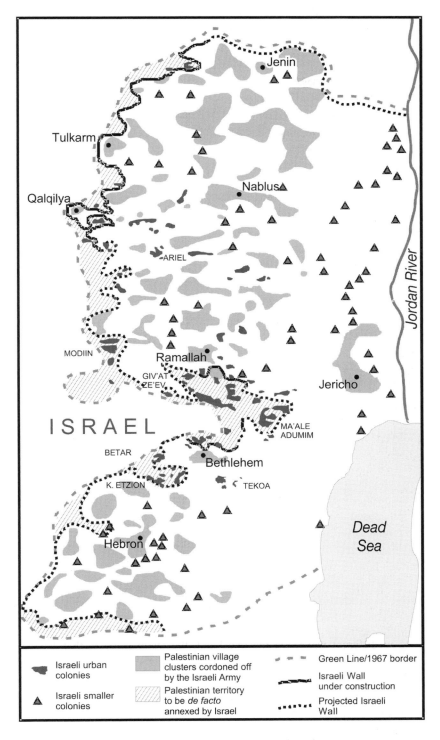

Figure 5.5 Israel–Palestine II: Palestinian villages and Israeli settlements.

emphasizes that their security needs are being met by the construction of the wall; it is seen as a barrier to prevent suicide bombers and other terrorists entering Israel and killing their citizens. There is, of course, some ground for their stance. However, the wall has been imposed upon the Palestinian population with no consultation and has amounted, in some cases, to a "land-grab," as some Palestinian villages have found themselves on the Israeli side of the wall.

The construction of walls and territorial areas is only part of the construction of territory that is the ongoing means of the Israelis' control of the Palestinians. The architectural scholar Eyal Weizman (2007) has described how the built landscape at the scale of buildings, checkpoints, and road routes restricts and defines movement and access in such a way that the occupation of the West Bank is facilitated by, as well as the reason for, the construction of Israeli settlements. Weizman's work forces us to consider how the construction of territory for political reasons is more than overt political boundaries, but a matter of everyday landscapes that allow for the observation and control of the weak by the powerful. Technology, architecture, and ideology intersect to create spaces that enable some at the expense of others.

The issue of human rights remains when considering Israel's treatment of the Palestinian population. The destruction of Palestinian homes and olive groves that have stood for generations has been a constant focus of complaint by the Palestinians and human rights organizations. An Amnesty International report (2004) notes how, in the three or four years leading up to 2004, 3,000 homes, and hundreds of public and commercial buildings, plus vast areas of agricultural land, were destroyed in the Occupied Territories. It is estimated that around 20,000 people were displaced (Falah and Flint, 2004).

The recent situation has seen the increasing role and profile of Hamas, an Islamist group that has challenged the secular Fatah movement as being the leader of the Palestinian people. Following elections in January 2006 in which Hamas defeated Fatah, the US, the EU, and Israel enacted severe economic sanctions, called by some a "blockade" on the Gaza Strip causing severe hardship to the population. The blockade continued through 2007 and there was fighting between Hamas and Fatah forces. In June 2008 Hamas agreed to a ceasefire with Israel but this broke down after rocket attacks from Gaza into Israel, which caused anxiety within the Israeli population though casualties were very low. In December 2008 Israel launched a wave of airstrikes on Gaza over a three-week period, and then Israeli troops entered Gaza. It is estimated that over 1,400 Palestinians and thirteen Israelis were killed in this phase of the conflict.

Since the violence of December 2008 and January 2009 there has been relative peace, but no steps toward resolution. President Obama's attempts to resurrect a peace process have been rebuffed by Israel. Despite the disapproval of the US government, Israel continues to build in the West Bank.

What are the sticking points or major barriers to peace in this conflict? Four main issues stand out:

- dispute over the control of Jerusalem; Palestinians have been increasingly excluded;
- Right of Return for refugees;
- a sovereign Palestinian state, but one that has territorial congruity and meaningful sovereignty from Israeli demands;
- unequivocal recognition of Israel and its right to exist in peace with its neighbors.

The construction of the security wall, the Israeli settlement of the West Bank, the continuing Israeli control of Jerusalem, continued terrorist efforts by Palestinian militants, the dubious efficacy of the Palestinian Authority to harness the militants, the increasing role of Hamas, unresolved boundary disputes with Syria, and doubts about the US's ability to play the role of "honest broker" conspire to make the path to peace most challenging.

The case study illustrates some important points about the geopolitics of boundaries. Demarcation and establishment through war are unlikely to make a peaceful boundary. The political goodwill necessary for the construction of peaceful boundary relations is nigh impossible to cultivate when there is gross disparity of power between the geopolitical agents. Boundary conflicts are not merely the product of the local geopolitical codes of neighboring states but are often the product of the geopolitical codes of other states. Boundary disputes are inseparable from the politics of nationalism, and so identity plays a central role – including particular interpretations of history. Identity and control of movement were seen to be the key issues in the Israel–Palestine conflict, but these central issues may also be seen in an opposite light, as the sources for peaceful cross-boundary interaction.

The geopolitics of making peaceful boundaries

Boundaries are the focus for a variety of geopolitical disputes, but does the concentration upon the geographic line in the sand, an absolute marker of national identity and state sovereignty, provoke conflict? Boundaries create an absolute world of being either completely within a particular nation-state, or completely outside of it. There is no grey area in this geopolitical vision; the resource is either Hypothetica's or not, an individual is either a Hypothetican or not. Some argue that a more productive approach is to emphasize the geopolitics of borders rather than boundaries. Reflection upon borders and borderlands may result in trans-boundary interactions that allow for mutual control and utilization of resources and joint economic activities.

To make a peaceful boundary, political goodwill between neighbors is fundamental: mutual trust and shared goals are the basis for cooperation (Newman, 2005, p. 336). Specifically, the following conditions are necessary to facilitate trans-boundary interaction (Newman, 2005, p. 337):

1   Territorial questions are settled. There is no dispute over where the boundary has been established and how it has been demarcated.
2   Trans-boundary interaction within the law is easy. The boundary facilitates flows (tourists and labor migrants, for example) between neighboring countries rather than preventing them.
3   The boundary provides a sense of security. Rather than being seen as a source of potential conflict, the boundary is seen as a sign of strength as commuting and joint economic projects enhance well-being and eradicate concerns of potential warfare.
4   Joint resource exploitation is possible. The basis of the peaceful boundary is mutual economic growth through interaction. For example, shared lakes,

rivers, and aquifers may be managed jointly. Other examples are the "peace parks" or "free enterprise zones" that minimize the existence of the boundary by creating tariff free international trade. The boundary as the enclosure of state-imposed taxation is loosened by these zones.

5   Local administration is coordinated. Emergency services and transportation logistics are examples of how local governments in neighboring states can create functional integrated areas that straddle an international boundary.

In introducing trans-boundary cooperation, the focus is upon how two states interact politically for economic purposes. The coordination of local administration facilitates interaction, with the main goal being economic gain: increased trade, commuting to work across a political boundary, or jointly harvesting timber or fishing a lake, for example. The assumption is that the increased economic efficiency will strengthen the legitimacy of the separate states. However, cooperation may provoke other questions and concerns. What about issues of identity, if the role of the boundary in delimiting national identity diminishes, and what impact does this have on the way individuals in the borderland identify themselves?

Borderlands

Interest in the cultural question of identity has focused attention upon borderlands (Martínez, 1994). The borderland is a trans-boundary region that shares common cultural traits, producing a geographic region of identity that is different from the two contiguous national identities. The borderland trans-boundary identity challenges the ideology that state boundaries encompass a national identity (Appadurai, 1991). Instead, borderlands require consideration, on the one hand, of the fractured nature of national identities, and, on the other hand, of the commonalities (rather than differences) across national groups.

There are five key processes that shape a borderland (Martínez, 1994):

1   Trans-nationalism: Borderlands are influenced by, and sometimes share the values, ideas, customs, and traditions of, their counterparts across the boundary line. Hence, the ideological unity of national culture is challenged, as is the idea of state boundaries acting as the "containers" of national identity.

2   Otherness: The borderland is culturally different from the majority of both of the states' populations it is part of. The majority of the two states' populations view the inhabitants of their border region and, perhaps to a lesser extent, the whole borderland as exhibiting different cultural traits.

3   Separateness: The cultural otherness of the border and the borderland can result in an ideological and functional separateness from the rest of the state. Separateness may manifest itself in discrimination toward the border culture in education and the media, possibly with manifestation in government employment. In addition, the states' infrastructure may be relatively inefficient in the border region. Either alone, or in combination, cultural

and functional separateness can make the two borders peripheral to their respective states. In light of this status, shared cultural traits across the boundary may foster solidarity and cooperation.

4   Areas of cultural accommodation: Peripheral status and discrimination within their respective states may encourage the residents of a borderland to forge a sense of solidarity that transcends ethnic differences. The "them" and "us" dichotomy that a state boundary fosters can be undermined as collective identities that cross a state boundary and challenge national homogeneity are created.

5   Places of international accommodation: Functional cooperation and cultural fusion can foster borderlands as zones of international cooperation, especially if economic integration and joint security and military operations have muddied the notion of state sovereignty being a singular enterprise that stops at the boundary. Instead, responsibility for security and economic growth is shared by two states, and its scope is no longer bounded by what has been understood as the geographical limits of the state.

The reason why scholars have increasingly focused upon borderlands is the role they play in creating geographies of identity and economic cooperation that are not based upon state boundaries and their ideological overlay with the pattern of national identity. If the boundary is key in establishing a state and nation, borderlands could play a role in challenging states and nations.

The geopolitics of identity, of which borderlands are one example, is challenging the importance of the hyphen in nation-state (Appadurai, 1991). The ideology of the nation-state asserts that all those within the boundaries of a state are members of a common nation. Going back to the chapter on nations and nationalism, we saw that national separatist movements are practicing a geopolitics based on the idea that a particular state contains more than one national identity, and minority nations have a right to their own state. Appadurai alludes to a different geography: the geography of cultural groups is not a mosaic of nations that can be given territorial expressions as nation-states. Instead, cultural groups are tied together across the globe in networks of migration and cultural association that are played out over and in the boundaries of states. Networks of cultural association intersect state boundaries. Territorial manifestations of identity are sub-national, connected to regions and localities within states. As ethnic groups settle in particular parts of a state they may construct a regional identity. Alternatively, the group may assimilate and move within the state, which reduces geographic concentration over time.

Case study: global geopolitical codes and the establishment of the North Korea–South Korea boundary

Korea's recorded history dates back to 57 BCE, dominated by periods of subservience to the Chinese Empire. However, this changed in dramatic form at the end of the Sino-Japanese war of 1894–5 when both Japan and China recognized Korea's complete independence. In the wake of Japan's victory, conflicting Japanese and Russian interests in Korea led to the Russo-Japanese war of 1904–5. Japan's victory stunned the Western world, where dominant racist ideology had made an Asian victory over a European state unthinkable. The final settlement to end the war was brokered with the aid of the President of the United States, Theodore Roosevelt. Japan was permitted to occupy Korea through the Treaty of Portsmouth of September 1905. By 1910 Korea was forcibly annexed and incorporated into the Japanese empire (Collins, 1969, p. 25).

Korea was administered as a colony within the Japanese empire between August 22, 1910 and September 6, 1945. Facing both Chinese and Soviet attempts to exert influence in Northeast Asia, Japan became increasingly anxious to develop a regional geopolitical code. Korea was a key part of Japan's expansion into mainland Asia. In a quid pro quo between global and regional geopolitical codes, the United States and Britain were willing to give Japan free reign in Korea in exchange for Japanese recognition of their interests in Asia and the Pacific. Japan justified its occupation by portraying it as a "civilizing mission" of modernization (Hoare and Pares, 1999, p. 69). However, the objective of these developments was to turn Korea into a dependable and productive part of the Japanese empire (Hoare and Pares, 1999, p. 69). Furthermore, the occupation was brutal, fostering an animosity toward Japan that remains, to some extent, today.

The animosity bred a nationalist geopolitical code of resistance. At the beginning of the twentieth century, camps were established to train a military force to resist the Japanese occupation, while other groups tried to gain assistance for the independence of Korea in a more diplomatic way, lobbying foreign governments. For example, Syngman Rhee, later to be the first president of South Korea, established the Korean National Association in Hawaii in 1909 (Eckert et al., 1990).

In the wake of World War I, the United States began to disseminate a global program of national self-determination. Koreans interpreted the context as one in which the major powers would be sympathetic to their own goals of ending the Japanese occupation. On March 1, 1919, a peaceful uprising burst out when a Declaration of Independence, prepared primarily by religious groups, was read out in Seoul. In the wake of fierce suppression many Korean nationalists fled to China. A Korean provisional government was established in Shanghai in April 1919. However, the Korean exiles were very scattered and divided politically. These divisions were reflections of different perspectives on how to bring the Japanese domination of Korea to an end, as well as various ideologies (Hoare and Pares, 1999, p. 24). In other words, though the geopolitical goal of the groups was common, the means to achieve it were disputed.

The establishment of the Soviet Union had promoted the diffusion of social revolutionary thought. Socialism spread first among Korean exiles in the Russian Far East, Siberia, and China, and then among Korean students in Japan, attracted by its combination of social change and national liberation. The different groups of exiles continued to clash, sometimes violently, over ideological differences. The Korean nationalist movement was too weak to end Japanese occupation. Instead, Japan was driven out of Korea in the wake of its defeat in World War II and the dissolution of its empire. Differences amongst Koreans remained unresolved (Hoare and Pares, 1999, p. 24).

Almost immediately, efforts were made to form a Korean government with its headquarters in Seoul. Initially named the Committee for the Preparation of Korean

Independence, on September 6, 1945 the government changed its name to the Korean People's Republic (Cumings, 1997, p.185). Soviet troops had been fighting the Japanese in Korea since August 8, 1945. They gave "permission" for US troops to enter Korea further south than Seoul, while supporting the Korean People's Republic (Cumings, 1997, p. 186). As part of the redefinition of the US geopolitical code at the beginning of what came to be known as the Cold War, it did not recognize the republic the Soviet Union had helped create. In a move that presaged the division of Korea, the US chose instead to support the nationalist exiles and the few conservative politicians within Korea who comprised the Korean Democratic Party (KDP). Within a context of competition between two external powers, Koreans made political choices, and within a matter of months Korea was divided into Socialist and Capitalist political allegiances with, virtually, a North and South geographic expression respectively (Cumings, 1997, p. 186).

The subsequent division of Korea had no historical or political basis. "If any East Asian country should have been divided it was Japan," writes Bruce Cumings (1997, p. 186), given its role as aggressor in World War II. The thirty-eighth parallel that was originally chosen to divide Korea had no prior meaning for Koreans, but now it is central to their lives (1997, p.186). Instead, the demarcation of the boundary was a product of the geopolitical codes of the Soviet Union and the US. The thirty-eighth parallel was first established as the dividing line of Korea on August 10, 1945 by Dean Rusk and Charles H. Bonesteel, two American colonels who had been instructed to do so by John J. McCloy of the State-War-Navy Coordinating Committee (SWNCC) (Cumings, 1997, pp. 186–7). Their rationale was to include Seoul, the capital city, within the American Zone. Surprisingly, the Soviets accepted the division. Unbeknown to the Americans, the Soviets and the Japanese had themselves discussed dividing Korea into spheres of influence at the thirty-eighth parallel. Rusk confessed many years later that, "Had we known that, we all most surely would have chosen another line of demarcation" (Oberdorfer, 2001, p. 6). The decision was made without consulting any Koreans (Cumings, 1997, p. 187).

On August 15, 1948, the US-backed Republic of Korea was officially proclaimed and on September 9 the Soviet-backed Democratic People's Republic of Korea was proclaimed in the North. The Soviet Union chose Kim Il Sung (born Kim Song Ju), a 33-year-old Korean guerilla commander who had initially fought the Japanese in China but had spent the last years of World War II in Manchurian training camps commanded by the Soviet army, to lead its regime in the North. In the South the US chose 70-year-old Syngman Rhee as the first Korean president. He was a product of contacts with the US, and had obtained degrees from George Washington University, Harvard, and Princeton. Both leaders felt they were destined to reunite their country.

After the creation of these regimes both Soviet and US troops left the peninsula in 1948 and 1949, respectively. Just a matter of weeks after the US troop withdrawal, civil war broke out in the peninsula. On June 25, 1950, North Korea, with the support of the Soviet Union and China, invaded the South in an effort to reunify the country by force. The invasion was challenged and repulsed by the forces of the United States, South Korea, and fifteen other states under the flag of the UN. The United States pledged support for South Korea against North Korea and sought legitimacy through the UN. In resolution

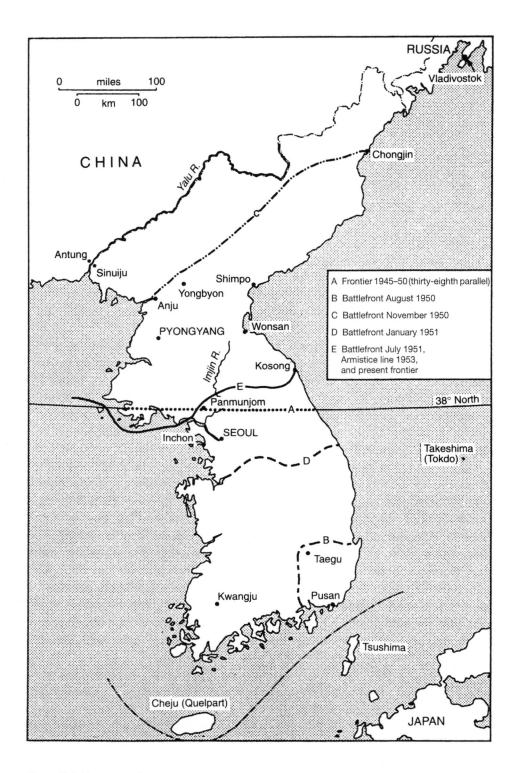

Figure 5.6 Korean peninsula.

No. 83 of June 27, 1950, the United Nations Security Council recommended that the member states of the UN should provide assistance to South Korea. The UN created a "unified command" (Hoare and Pares, 1999, p. 194), and asked the US to name a commander for it. President Harry S. Truman appointed General Douglas MacArthur. MacArthur had also been the Supreme Commander for the Allied Powers in Japan (Hoare and Pares, 1999, p. 194). The distribution of ground forces for the United Nations Command was 50.3 percent US, 40.1 percent South Korean, and 9.6 percent others. The United States provided the majority of naval and air force units.

The invasion came after Kim Il Sung had repeatedly requested authorization from Joseph Stalin, the Soviet leader. Stalin eventually approved the war plan due to what he called the "changed international situation." What this means remains debated. Possible reasons are the victory of Mao's Communist Party in China, the development of the Soviet Union's atomic bomb, the withdrawal of US forces from South Korea, or a statement by Secretary of State Dean Acheson excluding South Korea from the US defense perimeter; all of which occurred in 1949 or early 1950 (Oberdorfer, 2001, p. 9). The Korean War was a proxy war, a war fought between the superpowers through their allies rather than direct conflict between the Soviet Union and the US. It lasted from 1950 to 1953, and fortunes swung back and forth until an armistice agreement between North Korea, China, the US, and the UN was signed on July 27, 1953 (Hoare and Pares, 1999, pp. 3–4). The nature of the agreement means that the war is still unresolved; no final treaty has been signed. It is estimated that 900,000 Chinese and 520,000 North Korean soldiers were killed or wounded, as were 400,000 UN Command troops, nearly two-thirds of them South Koreans; 36,000 US soldiers were killed (Oberdorfer, 2001, pp. 9–10).

The end of the fighting resulted in the demarcation of a boundary close to the 38th parallel, a process initiated and defined by foreign countries. To this day the very limited flows across the boundary are controlled with the assistance of US soldiers stationed in South Korea, and the boundary is highly militarized. On the South Korean side, minefields line the roads, bridges are fortified, and checkpoints and gun emplacements are visible. The North Korean border is inaccessible. The war is, technically, still going on, and even today there are still fears in both Koreas that the fighting could break out at any moment.

In the aftermath of the fighting, the Rhee regime in the South became increasingly dictatorial and corrupt, until it was deposed in 1960 by a student-led revolt. There were numerous coups and assassinations in South Korea until its government finally seemed to normalize in the late 1980s. In the North, Kim Il Sung systematically purged his political opponents, creating a highly centralized system that accorded him unlimited power and generated a formidable cult of personality (Oberdorfer, 2001, pp. 10–11). Kim Il Sung was in power for nearly five decades, and died of a heart attack in 1994, and was succeeded by his son Kim Jong Il.

In the late 1990s North Korea experienced terrible famines, killing approximately 2 million people and devastating life in North Korea (CNN, 1998). When Kim Dae Jung took office as president of South Korea in 1998, South Korea changed its geopolitical code toward North Korea from a hard-line policy from the Cold War to an engagement policy known as the "sunshine" policy (Heo and Hyun, 2003, p. 89). The US generally supported the "sunshine" policy during President Clinton's administration, and sought to negotiate an end to North Korea's development of nuclear weapons and long-range missiles (Heo and Hyun, 2003, p. 89). The US evaluated the politics of easing tensions over the Korean boundary as a means to advance its global geopolitical code. After George W. Bush took office the overall picture between the three countries changed, as the US defined a hard-line policy toward North Korea. President Bush included North Korea in his "Axis of Evil" defined in his 2002 State of the Union speech, and increasing focus has been placed upon North Korea's capabilities to build nuclear weapons and long-range missiles. Such statements by the US have hardened the attitude of North Korea and put the brake on negotiations aiming at increasing some flows of goods and people across the boundary.

The situation changed dramatically in the 2000s with the development of a North Korean nuclear weapons program. In 2003 North Korea withdrew from the Nuclear Non-proliferation Treaty (NPT), prompting unprecedented talks between China and the US. With North Korea's declarations that is has enough nuclear material to make up to six bombs, a series of talks between North Korea, South Korea, China, Japan, Russia, and the US (the "six party talks") were held. Though North Korea formally pulled out of these talks in June 2004, the September talks appeared to produce an agreement that North Korea would give up its nuclear weapons program in return for aid and security guarantees; an additional demand of a civilian nuclear reactor came later.

In 2006 the nuclear brinkmanship escalated as North Korea first tested a long-range missile and then, in October, claimed to have tested its first nuclear weapon. Though this is a sign of North Korean nuclear power, the fragility of the state is exposed through its appeal for aid following another round of famine. In 2007 and 2008 there were signs of warming of relationships; the South and North Korean Presidents met and (as a sign of South Korea's so-called "sunshine policy") pledged to initiate talks about formally ending the Korean War; the New York Philharmonic orchestra performed in Pyongyang. But with the election of South Korea's new conservative President Lee Myung-bak tensions soon appeared. Military incidents and shootings occurred around the ceasefire line (the boundary between North and South Korea). In March 2010 the South Korean warship *Cheonan* was sunk, allegedly by a North Korean attack. Tensions rose as the US imposed sanctions and conducted joint military exercises with South Korea. Concern about North Korea's nuclear program continues, and in November 2010 North Korea fired shells into South Korean territory, killing two soldiers. In the background are reports of new levels of agricultural failure and famine in North Korea.

The story of the Korean peninsula is one of a militarized boundary that is virtually closed to movement. The boundary is a product of external geopolitical influence that reached its most violent form to date in the Korean War. Its establishment, demarcation, and control were a component of the Cold War. More recent attempts by Koreans to change the boundary regime have been hindered within a new geopolitical context that has focused US attention upon North Korea's nuclear capabilities. To date, the nature of the Korean boundary is very much a product of geopolitics operating at the global scale, making any intermittent agency by the two Korean states toward a more open boundary problematic.

Boundaries and geopolitical codes

Boundaries and borders are an integral component of a state's geopolitical code. The legitimacy and tenure of a government depends upon its ability to maintain boundaries from external threat. The identity of a nation depends upon the effective use of the boundary in maintaining a sense of geopolitical "order" which is the maintenance of a particular domestic politics in the face of "outside" threats. The separation of a domestic "inside" from an "outside" realm of foreign policy has always been a fiction, but, arguably, this is increasingly so in the wake of intensified economic integration of the globe and related cultural and migratory flows. Nevertheless, governments feel the need to maintain the distinction in their policy and rhetoric. To take the British example again, the goal of introducing identity cards for all British citizens was built upon a geographic understanding of the world: the inside group had rights and privileges that needed to be protected from the undeserving "outsiders." However, such an "inside" and "outside" distinction denies Britain's historic and contemporary connections across the globe. The politics and demographics of the United Kingdom, as with other countries, are the product of colonial expeditions abroad that have resulted in dramatic social changes "back home." For example, increased government control of the economy in the first half of the twentieth century was a function of the two World Wars, and patterns of immigration and citizenship through the twentieth century reflect the construction of Empire and Commonwealth.

At the end of Chapter 2 we discussed the establishment of the US's War on Terror as a geopolitical code. The emphasis in that discussion was the creation of US influence beyond its own boundaries, or a global projection of power within the role of world leadership. However, despite its global role the government of the US must still maintain its legitimacy by defending its boundaries. The terrorist attacks of September 11, 2001 violated the boundaries of the US like never before; to maintain its legitimacy (and get re-elected) President George W. Bush's administration had to convince the American people the boundaries would be secure in the future. Part of the rhetoric of boundary defense appeared within the justification of the military invasion of Afghanistan, as shown in Chapter 3, but the notion of boundary security, the protection of sovereign territory, was prominent too.

For example, in the following quotes (taken from Flint, 2004) the defense of a national territory through increased control and monitoring of movements through the US boundary is emphasized. In the words of Tom Ridge, about to become Secretary of the newly developed Department of Homeland Security at the time: "It's one war, but there are two fronts. There's a battlefield outside this country and there's a war and a battlefield inside this country." (Governor Tom Ridge, October 22, 2001. Excerpts from News Conference on Anthrax in Postal Workers.www.nytimes.com/2001/10/22/national/22CND-EXCE.html.Accessed January 10, 2002).

In other words, an attempt was being made to define two separate spheres of geopolitics, internal and external. An invigorated policing of the boundary was the geopolitical practice that was invoked to make this possible:

I'd like to note that the INS has been and continues to be a very vital player in this war on terrorism, in this investigation, as well as the ongoing process of protecting the American people from what we see as the forces of evil.

(Jim Ziglar, commissioner of the Immigration and
Naturalization Service, AG Outlines Foreign Terrorist Tracking
Task Force October 31, 2001. www.justice.gov/ag/speeches/
2001/agcrisisremarks10_31.htm.Accessed January 12, 2002.)

Put more simply: "You know, we used to think of America the beautiful, fortress America, trusting America. And we find that perhaps we've trusted too much." (NBC Nightly news Tom Brokaw interview with Tom Ridge, October 16, 2001. www.msnbc. com/news/643696.asp. Accessed January 12, 2002).

In the wake of the terrorist attacks of September 2001, the US faced a dilemma: it felt the need to intensify its global reach and influence, in an increasingly militarized form, at the same time that it felt the need to enhance the impervious nature of its own boundary (Flint, 2004). As its actions blurred a sense of division between domestic security and global reach, the boundary became an increasingly significant component of US geopolitical practice, rhetoric, and popular identity. Such blurring of boundary construction has been ongoing since the immediate wake of the September 11, 2001 terrorist attacks. As we saw in the first box in this chapter, boundary construction involves the everyday monitoring of law-abiding individuals – even in their own homes. Contemporary boundary construction is an essential component of a state's geopolitical code, but it is as much a "domestic" practice as an act of foreign policy.

Territoriality of the ocean and territorial disputes

A focus on the geopolitics of territory should include consideration of maritime disputes. The oceans have their own territoriality that is framed around the distinction between national and international waters. In 1609 Dutch jurist Hugo Grotius advanced the principle of *mare liberum*; the now taken for granted belief that the sea is international territory allowing anyone free access for the purposes of peaceful trade. This policy was established to facilitate global trade. But not all the ocean is deemed international. A substantial portion is claimed by coastal states through their declaration of control over the parts of the oceans "near" to their coastlines. "Near" is currently defined by the legal term "exclusive economic zone" (EEZ). The EEZ may be a maximum of 200 nautical miles from the coast and within it the coastal state lays claim to fishing rights and rights to exploit minerals under the seabed (Glassner and Fahrer, 2004, p. 453).

The intersection of state sovereignty and claims to the ocean's resources explains why states dispute ownership of small islands and outcrops of rocks. These pieces of territory may appear worthless but are a segment of national territory that defines the extent of the EEZ, or "national waters," and the fish and mineral resources within and below the compartmentalized sea. Many such disputes exist at the moment, but the oceans of East Asia and Northeast Asia contain some of the most interesting and potentially problematic.

Some of these disputes stem from the definition of boundaries at the end of World War II as the Japanese empire was defeated and the Soviet Union extended and established its presence in the region.

Three conflicts are consistently in the news. The Paracel and Spratly Islands are small coral outcrops in the South China Sea and straddle the Pacific and Indian Oceans. The islands themselves are barely above sea level and practically uninhabitable, but the region is of interest because of the oil and natural gas reserves under the seabed. Also, the islands are in a key strategic position in the important sea route between the Middle East and the oil-consuming countries of East Asia, notably China and Japan. China and Vietnam fought over these islands in 1976 and in 1988, when Vietnamese boats were sunk and over seventy sailors lost their lives. The dispute continues and has become increasingly tense since 1993 when China released a map depicting the nearby Natuna Islands within their national waters – an area that contains some natural gas fields currently claimed by the Philippines and Malaysia. Other maritime boundaries in the region are disputed between Vietnam and China and also between Thailand and Cambodia, preventing oil exploration. For further details see www.globalsecurity.org/military/world/war/spratly.htm.

The Kuril Islands stretch from the southern point of the Kamchatka peninsula in Russia to the northern tip of Hokkaido island (Japan). The southernmost Kuril Islands were occupied by the Soviet military at the end of World War II and remain under Russian authority, but are claimed by Japan, who call them the Northern territories. Part of the geopolitical struggle over the islands has appeared as a scientific tactic by the Japanese, who claim that some of the islands under Russian control are not actually part of the Kuril chain. The dispute with the Soviet Union prevented a formal peace agreement between the two countries at the end of World War II. Accommodation was finally obtained through the 1956 Soviet-Japanese Joint Declaration, but this agreement was resisted by the United States who, in the midst of the Cold War, wanted the Soviet occupation to be deemed illegitimate. The situation remained fairly stable for some decades after 1956, but (quite surprisingly) the dispute has still prevented formal closure of World War II between Russia and Japan. Recently, certain incidents have heightened tensions: especially visits to the region by Russian President Medvedev and Japanese school textbooks representing the islands as under Japanese sovereignty. However, there have also been some positive statements by the leadership of both countries indicating a desire to resolve the dispute.

South Korea and Japan are currently embroiled in a tense dispute, with a centuries-old history, over tiny islands that lie in the ocean between them. These islands are known as Dokdo (by the Koreans), and Takeshima (by the Japanese), and are also known as the Liancourt rocks. Both countries resort to historic tales to proclaim the legacy of their rightful sovereignty over the islands, as well as the use of biology and geology to give scientific legitimacy to their claims. The nationalist significance of the dispute lies in the fact that it was the first piece of territory annexed by the Japanese in their colonization of Korea in 1905; control that lasted until the end of World War II. Since the 1950s South Korea has staffed a lighthouse on the islands as a symbolic act of possession.

The intensity of the dispute increased after 2004 when exploration for oil and gas resources under the seabed surrounding the islands was initiated by the South Korean government. Japan has inflamed South Korea by sending its own geological survey team to the area. More information can be found at www.globalsecurity.org/military/world/war/liancourt.htm.

These three examples illustrate how territorial disputes over maritime boundaries are an intersection of material practices aimed at exploitation of natural resources and representations of the dispute that reference longstanding nationalist beliefs. Though the islands may be small they are the territorial manifestation and focal point of broad historical geopolitical processes that are the continuation of imperial projects and the prosecution of World War II. These disputes, and others not discussed in detail, require careful management so that they do not become catalysts of future wars. More information on these disputes, and any contemporary boundary dispute in the world, can be found at the website of the excellent International Boundaries Research Unit at Durham University: www.dur.ac.uk/ibru/.

Summary and segue

In this chapter we have focused on geopolitical practices that create territories as means of bounding or delineating political jurisdiction and identity. The key concept is territoriality (Sack, 1986). Though we have focused on international boundaries and the territoriality of states (as key geopolitical actors), processes of deterritorilization and reterritorialization are also important. Boundaries are the product and process of geopolitical agency. They are geographical features that are the manifestation of geopolitical actions, but they are also dynamic and contested geopolitical ideas and policies. A number of agents make boundaries the target of their geopolitical actions (governments, terrorists, nationalist groups) and boundaries are also the outcome of geopolitical processes operating at global, state, and sub-state scales. Actual and perceived boundaries, whether in existence or potentially established, provide the structure for geopolitical actions – whether it be the norms of international diplomacy or the terrorist actions of nationalist movements.

However, the emphasis upon flows in the academic discussions of borderlands or the policy imperatives of the Department of Homeland Security requires us to consider a very different geopolitics: networks and flows that cross political boundaries and so connect different places and territories. In the next chapter we explore the geopolitics of networks through a discussion of terrorism and social movements.

Having read this chapter you will be able to:

- Understand the concept of territoriality
- Identify geopolitical practices of deterritorialization and reterritorialization
- Understand how boundaries are an important part of the practice of geopolitics
- Identify the types of boundary conflicts within current affairs
- Understand why the establishment of boundaries is an important geopolitical practice
- Consider how geopolitical agency can undermine or change the roles boundaries play
- Consider the importance of maritime disputes over territorial demarcation

Further reading

Appadurai, A. (1991) "Global Ethnoscapes: Notes and Queries for a Transnational Anthropology," in R. G. Fox (ed.), *Recapturing Anthropology*, Santa Fe: School of American Research Press.

A provocative discussion of identity that emphasizes diasporas and multiple identities rather than nationalism.

Boyd, A. (1998) *An Atlas of World Affairs*, tenth edition, London: Routledge.

Though a little dated, this collection of short essays and maps is an extremely useful and accessible introduction to most of the world's conflicts.

Bregman, A. and El-Tahri, J. (2000) *Israel and the Arabs: An Eyewitness Account of War and Peace in the Middle East*, New York: TV Books.

It is practically impossible to recommend one book on any conflict, especially one as contested as this. But this book does an effective job of describing the main historic events in the conflict with the use of interesting interviews.

Donnan, H. and Wilson, T. M. (1999) *Borders: Frontiers of Identity, Nation and State*, Oxford: Berg.

An excellent survey and discussion on the literature addressing borders and boundaries.

Elden, S. (2009) *Terror and Territory: The Spatial Extent of Sovereignty.* Mineapolis: University of Minnesota Press.

A thought-provoking essay that provides a historic consideration of constructions of territory with particular pertinence to the way territory is being reworked within the War on Terror.

Martínez, O. J. (1994) *Border People: Life and Society in the U.S.–Mexico Borderlands*, Tucson: University of Arizona Press.

An in-depth study illustrating the nature of borderlands and their impact on boundaries.

Oberdorfer, D. (2001) *The Two Koreas.* Indianapolis: Basic Books.

A highly interesting and accessible introduction to the Korean peninsula conflict.

6

NETWORK GEOPOLITICS: SOCIAL MOVEMENTS AND TERRORISTS

In this chapter we will:

- Introduce the term meta-geography
- Discuss the geopolitics of globalization
- Identify the key attributes of transnational social movements
- Discuss the geography and history of peace movements
- Discuss the geopolitics of defining terrorism
- Identify the changing geography of terrorism over the past 100 years
- Identify the geography of contemporary religiously motivated terrorism
- Define the meta-geography of terrorist networks and counter-terrorism
- Introduce the geopolitics of cyber-warfare

Geopolitical thought and practice have been dominated by the state. The geopolitical codes of states are usually seen as being the most influential, and the classic geopolitical thinkers were advocates for the national security of their home country. The world political map is commonly identified as that of territorial nation-states. In the preceding chapters we have exposed some of these ideas as either partial or outright myths. Our task now is to recognize that the geopolitics of the world is one in which the construction of territorial entities, such as states, has always occurred in conjunction with the construction of networks to enable flows across the globe. The construction of networks and maintaining flows within them is no less a form of geopolitics than the construction of states and the practice of their geopolitical codes. In some instances states have actively participated in the construction of such networks, and in other instances they have resisted flows that they see as a threat. Often such actions take place at the very same time; such as contemporary actions to enable networks of finance and trade at the same time that terrorist networks and flows of migrants are identified as threats.

To gain a full understanding of contemporary geopolitics, in this chapter we will focus upon the geopolitics of networks. First, we will discuss the term meta-geography and its connection to the geopolitics of globalization. Then we show the necessity of a geopolitical perspective in understanding the substantive topics of transnational social movements, terrorism, and cyber-warfare. For each of these topics we concentrate upon how networks both challenge and are partially created by the state.

Box 6.1 The state-based politics of trans-national bailouts

In 2010 and 2011 the politics of debt-crisis hung over Europe. Portugal, Greece, and Ireland were the hardest-hit countries. In April 2011 rumors of the likelihood of Greece defaulting on its debt began to circulate. In this context, a new round of discussions between Portugal and the IMF and the EU took place. Portugal had already received a financial bailout, but its continuing dire economic situation required new discussions of transfers of money. Portugal's economy was evaluated by the behavior of international money markets: credit swaps and return rates on Portuguese bonds that were just one component of a swirl of transaction within global financial networks. It seemed that a state's future was being decided by actions within a global network.

However, the negotiations were complicated by something apparently quite traditional and territorial, an election in Finland. A party called the True Finns quadrupled the size of its vote and was likely to become part of a new ruling coalition. Finland has traditionally been pro-European Union, and hence likely to support a renewed bailout of Portugal, but the rise of the True Finns has changed the situation. The possibility of the Finnish parliament challenging an EU bailout put the ability for Greek and Irish debt plans to be restructured in doubt. The global markets responded accordingly, putting greater pressure on these troubled economies.

National politics and global financial markets intersected to illustrate that geopolitics is an intersection of territorialization and transnational flows.

Source: Julia Kollewe, "Portugal Bailout Could Be Affected by Election Gains for Anti-euro Finns," guardian.co.uk, April 18, 2011. www.guardian.co.uk/business/2011/apr/18/portugal-bailout-election-gains-anti-euro-true-finns. Accessed April 18, 2011.

Geopolitical globalization: a new meta-geography

The world has changed since the time of the classic geopoliticians. We now live within an era of globalization, a term used to describe the global economic, political, and social connections that shape our world. The state-centric view of the classic geopoliticians has been replaced by a contemporary focus upon globalization, or a geography of networks that cross boundaries and are expressions of power that cannot be tied to particular national interests. The networks cannot be connected simply to the interests of a particular country in the same way as, for example, Ratzel's vision reflected German interests and Mackinder's British goals. Geopolitics is not just the calculation of countries trying to expand or protect their territory and define a political sphere of influence; it is also about countries, businesses, and political groups making connections across the globe.

Meta-geography refers to the "spatial structures through which people order their knowledge of the world" (Lewis and Wigen, 1997, p. ix; Beaverstock *et al.*, 2000). Modern geopolitics, within the dominant framework of Anglo-American geography, has disseminated a meta-geography of the world as a mosaic of nation-states, despite the artificiality of these geographic units. For years, conflict between states was the focus of geopolitics: In other words, geopolitics was the sub-discipline that examined the power relations within the assumed meta-geography of nation-states. But the intensifying transnational networks of globalization are an emergent meta-geography in which flows of goods, money, and people across boundaries make banks, businesses, and groups of refugees, for example, important geopolitical actors. Political power is not just a matter of controlling territory, it is also a matter of controlling movement, or being able to construct networks to one's own advantage across political boundaries.

Let us contrast the geopolitics of globalization to the political vision of the classic geopoliticians. The economic concerns of, say, Mackinder and the German school were, for them, solvable through the exercise of political power by their own countries and by the extension of political boundaries. Countries were the most powerful geopolitical agents. In the era of globalization the geopolitical agency of countries has been limited as economic decisions must be made with reference to transnational economic organizations such as the IMF or WTO. Interest rates and currency values are set by the reactions of global markets and, in some cases, the IMF. Economic sovereignty is limited. In addition, the geopolitics of globalization has led to a dramatic increase in the number of geopolitical actors, especially non-governmental organizations and social movements.

Globalization is the contemporary manifestation of what has been a constant trend in world history; the ever closer integration of parts of the globe. There is a danger of thinking of globalization as a new manifestation of our age. Networks of communication (roads) have been essential in "tying together" state territories, the infrastructural power we introduced in Chapter 4. Networks of diplomatic relations were essential in maintaining

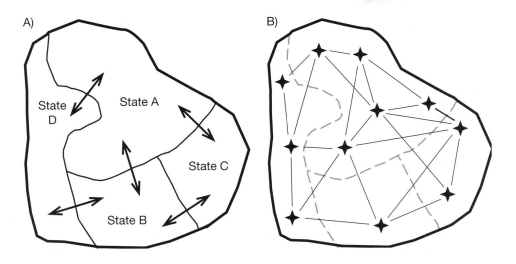

Figure 6.1 Meta-geography.

Box 6.2 Special Forces: the network power of the world leader

Networks of military power project the influence of the world leader across the globe. The increased role of US Special Forces since the invasion of Afghanistan in 2001 has been a mixture of covert military actions, but also "diplomatic" contacts with military forces across the globe. The former are militarized responses by the world leader to violent challenges; the latter are militarized attempts to maintain the US's global influence. The members of the Special Forces are highly trained and well-equipped killers, who have sought out the must dangerous form of modern combat. Ironically, much of their contemporary role consists of acting as "policeman," "diplomat," or, perhaps, "mayor" in conflict and post-conflict situations. Armed to the teeth, they are the visible expression of the global geopolitical code of the US in the "hottest" conflict spots across the globe.

For example, since 1981, Special Forces Sergeant Rick Turcotte has trained Fijian forces for peace-keeping missions, operated covertly in the Honduran jungle to help US-sponsored guerillas in Nicaragua, and supervised military training in Thailand, the Philippines, Malaysia, Indonesia, and Singapore (Priest, 2003, p. 124). The training missions fall within

> the bread-and-butter mission of Army Special Forces . . . "foreign internal defense," a concept refined in successive campaigns against communism but yet to be fully adapted for the post-Cold War period. This task calls for special forces to "organize, train, advise, and assist" a foreign military so that it can "free and protect its society from subversion, lawlessness, and insurgency," according to Field manual 31-20, "Doctrine for Special Forces Operations," issued in April 1990.
>
> (Priest, 2003, pp. 128–9)

This quote contains clues to the interaction between the agency of the world leader and the meta-geographies of nation-states and networks. The definitions of "subversion, lawlessness and insurgency" are made within the world leader's geopolitical code. "Society" is used here as another term for state; it is particular countries that are being assisted. However, the assistance is provided through a network of military units that are under less political supervision, within the US and abroad, than regular units (Priest, 2003, p. 139).

the new system of states that emerged after the Treaty of Westphalia in 1648. Societies have constantly explored beyond their immediate horizons, and networks of exploration date back thousands of years. In the history of geopolitics the networks of exploration established by Western states were essential in the practice of colonialism and the representation of "New Worlds." Networks of exploration enabled the establishment of territorial empires that, in turn, created flows of imperial trade and migration.

The current focus on globalization is a result of the intensification of these networks to such an extent that some see the construction of a global society, rather than an aggregation of national societies. Partially this intensification is the product of technological improvements that have allowed for quicker movement across longer distances for more and more people: from sailing ships through steam ships to jet passenger aircraft; from air mail through telephones to satellite technology; and from quite localized life experiences to global tourism and immigration. The construction of economic networks that have destabilized the sense of a "national economy" has meant the end of inter-state trade to movement of goods within intra-company networks that have plants and offices in numerous states. Financial markets are global in scale, hooked into the hubs of trading screens located in key financial centers (London, New York, Tokyo, Bahrain, etc.). States have tried to manage this integration by, for example, regulating domestic media markets to limit the number of outside broadcasts, and policing the flow of legal and illegal immigrants. However, the internet and satellite TV have made it increasingly harder for states to manage the flow of information, and often the ability of states to manage international migration is limited.

On the other hand, it would be wrong to think of a simple dichotomy between states and networks. In many ways, states have been active agents in promoting transnational networks. Free-trade and international investment provide just one example of states negotiating to allow for the movement of goods and money across their boundaries. Increasingly, states are giving decision-making power to trans-national organizations that have a direct impact upon the well-being of their population. For example, the WTO creates and adjudicates trade rules that have an impact upon jobs in particular states. The War on Terror has promoted a military network of cooperation between national police forces and armies across the globe (see Box 6.2).

Networks are inherently neither good nor bad; they are political constructs used for political ends. Though numerous forms of networks exist in the contemporary world and can be identified as being relevant to contemporary geopolitics, we will focus on two. First, we will discuss trans-national social movements and their attempts to forge "progressive" politics that transcends the scale of the state. Second, we will discuss terrorism, and the way it has changed over time to be identified as a trans-national threat to states.

Trans-national social movements

A social movement is a group of people organized, as groups of individuals and/or combinations of different groups, to pursue political goals in venues other than state institutions (such as voting). They may come together to promote the interests of certain groups (such as immigrants), or to focus upon a particular issue and goal (e.g. nuclear disarmament), or to challenge societal norms (such as sexuality). Beginning in the 1960s, two important changes in social movements have occurred. There has been a growth in trans-national social movements, or the organization of social movements to make connections across state boundaries. In addition, and related to the trans-nationalism of social movements, there is a change in the issues that are being addressed, with prominence given to environmentalism and peace movements.

The establishment of organized trans-national social movements is the result of four related changes or trends (Kriesberg, 1997): a growing trend toward democratization; increasing global integration in economic, political, and social spheres; converging and diffusing values that in turn bring people together in the name of a shared concern or issue that may be seen to be in opposition to other values and goals (e.g. environmentalism versus capitalism); and a proliferation of trans-national institutions that facilitate social organization beyond state boundaries. These trends are not necessarily new, but are seen as components of contemporary globalization that have intensified in recent decades. The trends should not be seen in isolation from each other, and are just as much trends creating globalization as they are outcomes of globalization.

The way that transnational social movements continue globalization trends is not just through providing an institutional infrastructure of communications and activity linking people in different states and different political agendas. It is also a matter of creating identities that focus on the trans-national or the global, rather than nationalism (Kriesberg, 1997, p. 14). But why is such identity formation a form of geopolitics? Kriesberg identifies five ways in which transnational social movements are able to alter the existing political landscape:

1   Mobilize support for particular policies.
2   Increase participation in the decision-making process.
3   Maintain the public's attention on critical issues.
4   Represent or frame the issues in a particular way.
5   Enact certain policies, or make such policies come about.

In combination, these five themes construct geopolitics as a product and process of mass activity, rather than the purview of elites, politicize many issues rather than traditional definitions of "national security," and create scales of political activity that transcend states and create global connections.

The geopolitics of transnational social movements identified by Kriesberg (1997) explicitly recognizes the importance of geographic scale. Smith (1997) identifies three scales that are targeted, though these should not be seen as being mutually exclusive: individual, state governments, and intergovernmental institutions. The individual is seen as a geopolitical scale in that their attitudes and behavior may be changed by the activities of the social movement. For example, eating preferences may be changed by environmental groups who highlight factory farming, or some campaigns ask consumers to boycott products from certain countries because of their political behaviors. Social movements also target states. For example, Greenpeace has sustained a long campaign against Japan because of its whaling practices. Anti-war protestors are usually targeting a particular state to change its geopolitical code. Finally, transnational social movements engage international organizations. The targets may be private companies (such as oil companies) or inter-governmental organizations such as the WTO.

So then what is an act of geopolitics? Smith (1997) identifies particular strategies that are directed at the three different scales. At the individual scale, a simple act such as holding a rally in which people are made aware of a particular issue and the impact of their actions is an act of geopolitics. Writing letters to state leaders and politicians is a geopolitical act targeting the state scale. At the scale of intergovernmental institutions, participating in the construction of an international convention would be an example in which a social movement works with formal institutions.

One good example of a social movement playing a role in a formal international institution is the politics behind the Third UN Conference on Law of the Sea (UNCLOS III). The conference ran from 1973 to 1982 and was the basis for the international laws of 1994 that established the territorial seas and economic resource zones we discussed as the territoriality of the sea in the previous chapter. UNCLOS III ran for so long because each country in the world had some particular concern or issue. In general, the richer countries wanted to ensure the global operation of their navies and fishing fleets, while the poorer countries wanted to make sure they has access to the ocean's resources off their coasts. Levering (1997) provides an interesting account of how the conference was facilitated by the actions of two concerned social movements: the Ocean Education Project and the United Methodist Law of the Sea Project, that collectively became known as the Neptune Group.

Both of the social movements in the Neptune Group had a commitment to world governance and came from a liberal Methodist background that promoted US engagement with the world to promote peace and international cooperation (Levering, 1997). The Neptune Group played a crucial role on UNCLOS, acting as "honest broker" between the negotiating states. Specifically, the Neptune Group was able to bring together experts and negotiators and was seen as a source of neutral and objective information. In the words of the Conference President, the Neptune Group:

> brought independent experts to meet with delegations, thus enabling us to have an independent source of information on technical issues. They assisted repre-sentatives from developing countries to narrow the technical gap between them and their counterparts from developed countries. They also provided us with opportunities to meet, away from the Conference, in a more relaxed atmosphere, to discuss some of the most difficult issues confronted by the Conference.
>
> (United Nations, 1982; quoted in Levering, 1997)

In sum, the simple acts of providing objective views and facilitating conversations enabled states to come to an agreement to produce a law of the sea that continues to fundamentally shape the territoriality of our planet.

Globalization and social movements

The anti-globalization movement provides a strong example of the diversity and fluidity of transnational social movements. It has no territorial center or stable agenda, but is continually changing its methods and goals as a result of interaction between the diverse groups of which it is comprised. Reflecting this lack of hierarchy and its eclecticism, the anti-globalization movement is also known as the Movement of Movements. The anti-globalization movement addresses a range of issues that range from ecological concerns to protests over economic neo-liberalism, to feminism. Such eclecticism produces no single and stable goal, leading to ridicule from those on the right of the political perspective, and criticism from those with a more traditional and state-centric left-wing agenda. However, its proponents claim that the fluidity of the movement is its very strength; enabling it to continually adjust to the dynamics of economic globalization and simultaneously showing the connections between issues of biodiversity, economic growth, democracy, and social marginalization. Furthermore, its lack of loyalty to a central organization prevents it from compromising on underlying beliefs; a multitude of movements will provide continual criticism, even of the movement itself. The number and diversity of movements creates connections across the globe to promote awareness of the way people in different places are connected by transnational economic and political networks. The movement has come together, though, in the World Social Forum conferences.

The eclectic nature of the World Social Forum (WSF) has been captured by an analysis of the way in which participants self-identify themselves with particular causes and actions. By surveying attendees of the 2005 WSF in Porto Alegre, Brazil, sociologists Christopher Chase-Dunn and Matheu Kaneshiro (2009) identified eighteen movements within the movement (see Table 6.1). A total of 560 respondents identified the types of groups they were most active in, and could list more than one type of group, to show the connection between group-types: human rights/anti-racism (12 percent), environmental (11 percent), alternative media/culture (10 percent), and peace (9 percent) were the movements with the most activity. By exploring the connectivity of cross-membership, a social network map of the interaction between different groups in the WSF can be created (Chase-Dunn and Kaneshiro, 2009) that shows which movements are most central, or form the hub, of the WSF's activity (Figure 6.2). This map indicates that human rights/anti-racism, environmental, and peace movements form a core of the WSF's activity and agenda. Also, this pattern was found to be stable through the 2007 WSF meeting (Chase-Dunn and Kaneshiro, 2009). (For more information regarding the content of past and future World Social Forums see www.nadir.org/nadir/initiativ/agp/free/wsf/. Accessed April 20, 2011.)

Table 6.1 Types of activism at 2005 World Social Forum (adapted from Chase-Dunn and Kaneshiro, 2009)

Type of group	Number of selections by respondents (total of 1,298 responses from 560 respondents)	Percentage of total selections
Anti-corporate	43	3
Anti-globalization	68	5
Human rights/anti-racism	161	12
Environmental	144	11
Fair trade	67	5
Peace	113	9
Queer rights	37	3
Feminist	66	5

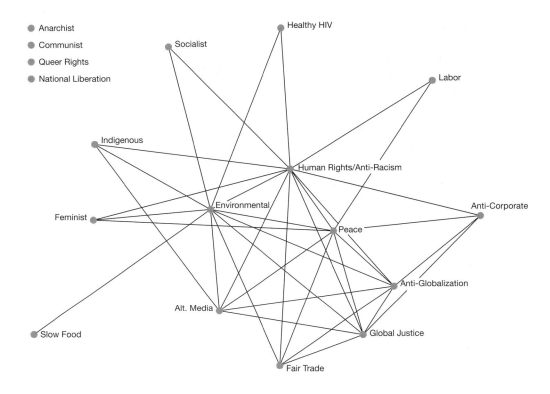

Figure 6.2 Group network connectivity in the World Social Forum.

The importance of peace movements to the WSF is an indication of resistance to the dominance of geopolitical codes of states that advocate and practice militarism and violent force. By looking at the geopolitics of social movements we can address how state-based militaristic geopolitics is being challenged by peace movements. The ability of contemporary social movements to connect individuals through transnational networks with the goal of challenging states is an example of a politics of scale that opposes the assumptions of classic geopolitics. Also, noting how peace movements are strongly tied to human rights, anti-racist, and environmental movements forces us to contemplate what is meant by "peace" and how peace activism may create a radically different global geopolitical imagination.

Geopolitics of peace movements

Writing within the context of the Cold War, George Konrad (1984) proposed the idea of antipolitics to challenge the nuclear militarism of the United States and the Soviet Union that had the potential to devastate the globe. Konrad believed that the pursuit of peace could not be left to states; they had a tendency toward militarizing issues and provoking

Box 6.3 Geopolitics of apology and forgiveness

Classic geopolitics is based upon mutual mistrust between states, and national identities that constantly look at past wars as the basis for continued militaristic foreign policies. Military history is all about past glories and failures that serve as "lessons" for continued preparation for war. An alternative geopolitics is one based upon apology or forgiveness that recognizes the legacy of past geopolitics and the benefits of recognizing their contemporary cost and impact.

British geographer Nick Megoran (2010) has studied the Reconciliation Walk, a grassroots US evangelical Christian project that retraced the route of the First Crusade in apology for it. The actions of this social movement must be understood within the context of President George W. Bush's reference to the war on Iraq against Saddam Hussein as a "crusade" and the fierce reaction that created within Muslim countries. The Reconciliation Walk aimed to address what its own organizers identified as "deep mutual hatred" between Christians and Muslims stemming from a geopolitical event, the Crusades, that took place around 900 years ago. The Walk attempted to follow the geography of the Crusades, stopping for reflection and interaction with the community at key sites, such as battlefields.

Megoran's study is part of a broader attempt to change the way geopolitics is conducted, as an academic and a practical exercise. He calls for a pacific geopolitics that would explore:

> the ways in which spatialising and ordering the world in imaginative geographies can contribute towards more harmonious relations between states and other human groupings. Pacific geopolitics is thus the study of how ways of thinking geographically about international relations can promote peaceful and mutually enriching human coexistence. Whereas critical geopolitics' focus has been a critique of war, pacific geopolitics would conduct theoretically informed empirical research on peace.
>
> (Megoran, 2010, p. 385)

The geopolitics of apology and forgiveness, in the form of the Reconciliation Walk, is a practical action of a social movement; though one with its own agenda to promote Christian fundamentalism (Megoran, 2010, p. 389). The very notion of forgiveness requires geopolitical thinking that connects people across time and space and builds mutual recognition of the costs of hatred and violence.

conflict. For Konrad, the solution was to reject state-based politics (voting for representatives, lobbying, etc.) completely and create alternative movements that crossed international boundaries to form communities of people seeking common goals and values. The antipolitics that Konrad called for countered the perspective of nationalism and its emphasis upon difference, or the Other; a tendency that was heightened by the antagonism of the Cold War. It is also easy to identify Konrad's vision as having come to some fruition

Figure 6.3 Reconciliation Walk participants praying in Jerusalem.

with the development of contemporary trans-national social movements and the central role of peace movements within the WSF that we have already discussed.

So what is peace and how have social movements organized to try and obtain it? The usual way to define peace is to distinguish between negative and positive versions of the condition. Negative peace is, simply, the absence of violence of all kinds (Galtung 1964; 1996, p. 31). Positive peace is better thought of as a process: a means to resolve conflicts peacefully and transform institutions and behaviors to promote justice and well-being (Galtung, 1996, p. 32). Positive peace endeavors to end violence in a sense that goes beyond simply stopping bouts of physical violence. Positive peace requires (1) identifying inequitable economic and social structures, transgressions of the natural environment, and attitudes of racism, homophobia, sexism, and religious fundamentalism, and (2) creating means to transform these structures and create dialogues of mutual understanding between individuals, states, and social groups.

Galtung (1996) realizes that the pursuit of positive peace requires consideration of different geographic scales. Adolf distinguishes three basic categories of peace that can be related to scale:

1   Individual Peace: How individuals become and stay at peace with themselves;
2   Social Peace: How groups become and stay at peace within themselves; and
3   Collective Peace: How groups become and stay at peace between each other.

(Adolf, 2009), p. 2)

If we relate this trifold category to the dominant actors of geopolitics and negative peace, the latter two categories can be thought of as peace within a state (lack of social

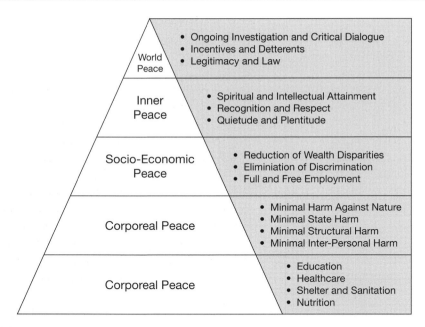

Figure 6.4 Peace pyramid.

disorder or civil war) and peace between states (lack of war between states). In terms of positive peace, Adolf's use of the word "become" is important: it forces us to consider institutions and behavior that constantly work to better the individual's sense of purpose and worth, the ability of states to better the life-chances of all social groups within their borders and the interaction between them, and inter-state cooperation that improves the well-being of all states. In other words, positive peace requires a progressive approach that creates a world based on a sense of collective identity and the mutual benefits of cooperation rather than merely the absence of fighting between states. The ideas and practices of the transnational social movements of the WSF certainly match these goals. Furthermore, the perspective, suggested by feminist geopoliticians, of focusing upon the individual in particular circumstances, the multiple forms of security they are seeking, and the need to focus upon interactions across the globe seems best suited to exploring ways to achieve these forms of peace.

If peace is to be obtained it will require constant activity to create and maintain it at all scales, from the individual to the global. Recognizing the multi-scale nature of the process, Adolf (2009) proposes a pyramid of peace (Figure 6.4). The structure of the pyramid reflects the three scales of trans-national social movement activity we introduced earlier: the individual, states, and intergovernmental institutions (Smith, 1997). As we shall see, the attainment of peace at all levels of the pyramid requires geopolitical agency, by individuals and groups, which create territorial entities *and* trans-national networks. Each of the levels of the pyramid has a number of components that must be enacted, lived, and constantly constructed to build and maintain peace.

Corporeal peace, or the well-being of the mind and body, requires nutrition, shelter and sanitation, healthcare, and education (Adolf, 2009, pp. 236–8). Sanctuarial peace may be interpreted as the construction of small-scale areas or territories that enable freedom from interpersonal harm, oppression, and attack from the state or other social groups. Socio-economic peace stems from the assumption that "how we live and work with each other (or not) as individuals and groups is a determining factor of whether peace is actualizable" (Adolf, 2009, p. 241). The components of socio-economic peace are full and free employment, elimination of discrimination, and reduction of wealth disparities. The next level of the pyramid is inner peace, composed of quietude and plenitude, recognition and respect for other social groups and political entities, and spiritual and intellectual attainment. Though this level may appear to be exclusively focused upon the individual scale, Adolf claims that it is relevant on individual, social, and collective levels (2009, p. 243). Individuals and groups require reflection, and the ability to respect and recognize others, and achieve a sense of good purpose. To put this another way, societies should not practice the "othering" we described as Orientalism in Chapter 4. The top level of the pyramid is world peace, which would involve not just peace between states but the actions of inter-governmental institutions and trans-national social movements to ensure the component parts: legitimacy and law, incentives and deterrents to maintain peace, and ongoing dialogue to understand and resolve differences.

The pyramid requires, amongst other things, creating areas of sanctuary (such as a peaceful home and community), a state with social and inner peace, and trans-national movements to constantly maintain respect and dialogue. Constructing and maintaining peace requires geopolitical agency at many scales and with the intention of creating territories and networks. Peace is a geographical and social-political process. The different components of the peace pyramid reflect the goals of different types of movements and their connectivity that we identified within the World Social Forum. Environmental movements engage issues of corporeal, sanctuarial, and social peace. Anti-racist/human rights movements engage the inner peace of the individual and the social, and the respect and dialogue necessary for social and world peace. Contemporary peace movements are not simply anti-war, but engage the other elements of the movement of movements identified by Chase-Dunn and Kaneshiro (2009). However, this has not always been the case. The following history of peace movements shows how the form of geopolitics undertaken by anti-war movements has changed to address peace more generally through a growing politics of transnational engagement.

Peace movements in time and space

As we noted at the very beginning of the book, geographers have been more active in creating geopolitical codes for states, and usually aggressive ones, than they have been in constructing peace. This is also true of the key classical thinkers of geography, with the notable exceptions of Kropotkin and Reclus. As with other social scientists, war has been a more intriguing topic for geographers than peace. The tide is turning to some degree (Mamadouh, 2005; Megoran, 2010; Kirsch and Flint, 2011b), and one topic of analysis is a geographic approach to peace movements and activism (Herb, 2005; Megoran, 2011; Koopman, forthcoming).

Herb (2005) identifies three periods of anti-war and peace activism, based upon the different geopolitical contexts within which the movements formed and the geographic scale at which they targeted their activity. Though Herb notes that there have been organized tendencies to promote peace throughout human history, he identifies the period after the Napoleonic wars as the origins of modern peace activity. The rise of nation-state politics (after the American and French revolutions) gave rise to a politically active citizenry. In conjunction with industrialized warfare, the citizenry was the basis for mass conscript armies that resulted in new levels of battlefield carnage. However, active citizens also organized to counter such state-based violence. In the mid-1800s the first organized peace groups appeared in the United States and Britain, including all-women groups, and in 1843 the first General Peace Convention was held in London (Herb, 2005, p. 351).

By 1900 over 400 peace societies existed in Europe and the United States, urged on by growing tensions on the European continent that, in hindsight, were the precursor to World War I. These efforts by citizens to promote peace had significant effects upon states. Herb notes the influence of peace movements in enabling Sweden's peaceful split from Norway in 1905, and also the Hague Conferences of 1899 and 1907 that outlawed inhumane weapons, such as poison gas, and the deliberate killing of civilians. Despite these successes, World War I started and unleashed levels of battlefield carnage that shocked the world. The politics of nationalism promoted the continuation of the war and made for a difficult political landscape for peace movements. States openly oppressed peace organizations, but still conscientious objectors (16,500 in Britain alone) and feminist movements pressed for a peaceful and internationalist agenda (Herb, 2005).

In between World Wars I and II the rise of fascist movements raised questions about the morality and effectiveness of peace movements, with some prominent voices saying that military means were necessary to counter such a political and social evil as Nazism (Herb, 2005). Though the world experienced nationalism again during the second global conflict, significant movements toward peace emerged. Notably Ghandi's anti-imperialist movement inspired people across the globe, and helped foster an understanding of peace beyond the negative sense of "not war." The rapid emergence of the Cold War after the end of World War II created a new environment hostile to peaceful inter-nationalism, especially during the aggressive and paranoid anti-Communism of the 1950s in the United States. In that context, any claims to be seeking peace and international engagement were easily labeled as being pro-Communist in the West and led to persecution.

Despite the constraints of political oppression, Herb identifies a second peak of peace activism in the late 1950s and 1960s. The geopolitical context was defined by the nuclear arms race and concern for the future of the planet. Nuclear bomb tests in the Pacific, the growing criticism of nuclear weapons by the scientific community, and the concern of countries outside of the bipolar alliances of the Cold War came together to launch anti-nuclear and peace campaigns. At this time political activities that can be seen as the roots for the contemporary organization of the WSF emerged. Notably, Women Strike for Peace was founded in Seattle in 1961, a grassroots movement that combined the goals of feminism and peace. There were some notable successes to the anti-nuclear movement. In 1967, twenty-four Latin American states signed the Treaty of Tlatelolco to declare a nuclear weapons free zone (Herb, 2005). The British Campaign for Nuclear Disarmament (CND) was founded in 1958, with similar organizations throughout Europe, and Canada,

New Zealand, and Japan. Anti-nuclear activism waned after 1964 with the passing of the Limited Test Ban Treaty; a partial success but enough to provide a sense of mission accomplished.

The third peak of peace activity occurred in the 1980s in a geopolitical context of heightened tensions in the Cold War (Herb, 2005). The 1979 NATO deployment of medium-range nuclear weapons in Europe changed the geography of the Cold War: countries in eastern and Western Europe, either side of the Iron Curtain, could foresee a limited nuclear war initiated by the superpowers that incinerated their cities while leaving the United States and the Soviet Union untouched. This new geopolitical context was the background for Konrad's (1984) antipolitics approach we introduced earlier. Concrete and practical manifestations of antipolitics included the intersection of environmentalist and anti-nuclear power movements, the women's movement and the increasing visibility of feminism, and questioning of Western consumer culture (Herb, 2005, p. 356). Though opponents of the movement, especially Western governments and mainstream media, painted the anti-nuclear movement as a puppet of a Communist conspiracy, public support was broad. In NATO countries surveys found support for the movement from

Box 6.4 A place for peace

Hiroshima is the site of the first nuclear bomb attack. It is sobering to reflect that nuclear weapons have been used by states in warfare. Furthermore, the two nuclear bombs that have been used in war were dropped by the United States at the end of World War II, and not by an international outcast or "rogue state"; and they were dropped with the intent to cause mass casualties amongst civilians rather than as limited or tactical weapons on the battlefield aimed at soldiers. Though two cities were the victims of US nuclear attack (Nagasaki being the other one), it is Hiroshima that plays an important role as a geographic site of remembrance for the horrors of nuclear war.

Hiroshima is an example of the construction of a place with a particular meaning and, hence, a role that expands beyond its simple location to encompass the globe. It is the site of the Peace Memorial Park and Peace Memorial Museum. The Park and Museum attract tourists from around the world and offers educational facilities aimed at furthering peace agendas. One of the activities of the museum is the construction of a network of universities teaching about the atomic bomb under the simple plea that "No one else should suffer what we did." The Peace and International Solidarity Promotion Division of the museum encourages universities across the world to teach courses on peace and the nuclear bomb. Hiroshima is a place acting as a key hub in an educational network that spans the globe and encourages peace.

Visit the site at www.pcf.city.hiroshima.jp/index_e2.html. Note the ways in which the language of the peace declarations, children's stories, and other exhibits use the special experience of Hiroshima and Nagasaki to create a global network and message.

between 55 and 81 percent of the population (Herb, 2005, p. 357). As with the previous peak of activism, success for the movement came in the form of a treaty, the 1987 Intermediate Nuclear Forces Treaty that stipulated the removal of short- and medium-range nuclear weapons from Europe.

Some general conclusions can be drawn from Herb's (2005) geographic interpretation of the history of peace movements. First, peaks of peace activism emerge within specific and different geopolitical contexts. Second, the social construction of scale is an integral part of peace activism, as individual and local actions are the building blocks for broader politics. Third, the scope of peace activism has changed over time, becoming increasingly global rather than national. Fourth, this movement toward a globalized peace movement goes hand-in-hand with an increasing breadth of political activity that has become the "movement of movements" we discussed earlier. Fifth, peace activity is grassroots based but requires a form of institutionalization allowing for a national and global impact. Sixth, and finally, peace activity by social movements has produced responses by states; the geopolitics of states has had to take account of coordinated demands for peace.

Transnational social movements are an example of the social construction of a geographic feature, a network, that is the means by which politics takes place and, at the same time, the result (or even goal) of that politics. Transnational social networks illustrate the interaction between states and networks. Another form of network that is much in the news is the terrorist network, especially the al-Qaeda network. The remainder of the chapter will further exemplify the geopolitics of a network meta-geography by focusing upon the agency of terrorists. Before exploring terrorism as networked geopolitics we must discuss the politics of defining terrorism and show how it has changed over time to become transnational.

Definitions of terrorism

The challenge to define terrorism is an impossible one for two reasons. First, terrorism has varied across history and geographical settings to make any one definition an inadequate description of the diversity of reasons and forms of terrorist activity (Crenshaw, 1981; Laqueur, 1987, pp. 149–50). Second, the definition of terrorism is in itself an act of politics: defining certain acts as terrorist acts makes certain forms of violence, political goals, and geopolitical agency illegitimate and so, in reverse, legitimates other forms of violence, politics, and agency. Defining a group as "terrorist" credits the form of violence that they inflict as being somehow "improper," "horrific," and "uncivilized." In calling these terms into question we by no means condone the murder of people in the name of politics. Instead, the purpose is to think about how the category "terrorist" helps us to accept other forms of violence as "proper," "reasonable," and "civilized."

Undefined terrorism

In Bruce Hoffman's (1998) accessible introduction to the topic of terrorism he takes great care to show the diversity of definitions of terrorism. Most telling is the table reproduced below (Table 6.2), which is a summary analysis of the predominance of particular terms

Box 6.5 War crimes?

In the documentary *The Fog of War*, former US Secretary of Defense Robert McNamara talks of his role as a strategist in the World War II firebombing of Japan that killed hundreds of thousands of civilians. In February 1945 one firebombing raid on the German city of Dresden destroyed 15 square kilometers of the inner city. Of the 28,410 houses in the area 24,861 were destroyed. Casualty estimates vary wildly, but recent scholarship puts the figure between 25,000 and 30,000; though some claim the total to be as high as 300,000. Overall Anglo-American bombing of Germany in World War II killed approximately 400,000 people, about nine times the 43,000 British citizens killed by German raids. Japan also suffered firebombing. Beginning in February 1945, the four conurbations of Tokyo, Nagoya, Osaka, and Kobe were targeted. One attack on Tokyo in March destroyed 41 square kilometers and killed an estimated 100,000 people.

In the documentary interview *The Fog of War*, McNamara says that if the US had lost the war he would likely have been tried as a war criminal for his part in the bombing. Was the shared Axis and Allies policy of bombing towns in World War II an act of terrorism? Give an answer now, and reconsider it in light of the discussion of definitions of terrorism below.

or concepts in 109 definitions of terrorism (Hoffman, 1998, p. 40). I draw attention to this figure precisely because of the lack of agreement or consistency that it illustrates. The most agreed upon aspect of terrorism is violence, which appeared in just 84 percent of the definitions – in other words, 16 percent of the definitions did not emphasize violence as an important component of terrorism!

The definition of terrorism is, at best, contested and, perhaps more fairly, unclear. However, we can still discern some important geographical elements of terrorism from the features listed in table 6.2. First is the symbolic nature of terrorist actions that promotes the targeting of particular places or buildings. The Alfred P. Murrah Federal Building in Oklahoma City was, for Timothy McVeigh and Terry Nichols, the local physical embodiment of the federal government that they viewed as an "occupying force" violating the freedoms of the American people. Less specifically, Palestinian terrorists target restaurants and buses in a brutal message that says that the public spaces of the state of Israel will never be safe until the rights of the Palestinian people for their own state are recognized (Falah and Flint, 2004).

Second, the goal of terrorism is to expand the geographic scope of a particular conflict in a manner that will, the terrorists hope, benefit their cause. Osama bin Laden has made the presence of US troops on the Saudi peninsula a matter that we must all consider, and something that becomes a part of electoral campaigns in Australia, Spain, Great Britain, the US, and beyond. The terrorist's perceived need to reach a broader audience, or expand the scope of "interested" or at least "implicated" parties, relates to the marginalization of some groups to the extent they resort to violence in order to place their situation on the

Figure 6.5 Dresden after Allied bombing.

Table 6.2 The problem of defining terrorism

	Definitional element	Frequency (%)
1	Violence, force	83.5
2	Political	65
3	Fear, terror emphasized	51
4	Threat	47
5	(Psychological) effects and (anticipated) reactions	41.5
6	Victim–target differentiation	37.5
7	Purposive, planned, systematic, organized action	32
8	Method of combat, strategy, tactic	30.5
9	Extranormality, in breach of accepted rules, without humanitarian constraints	30
10	Coercion, extortion, induction of compliance	28
11	Publicity aspect	21.5
12	Arbitrariness; impersonal, random character; indiscrimination	21
13	Civilians, noncombatants, neutrals, outsiders as victims	17.5
14	Intimidation	17
15	Innocence of victims emphasized	15.5
16	Group, movement, organization as perpetrator	14
17	Symbolic aspect, demonstration to others	13.5
18	Incalculability, unpredictability, unexpectedness of occurrence of violence	9
19	Clandestine, covert nature	9
20	Repetitiveness; serial or campaign character of violence	7
21	Criminal	6
22	Demands made on third parties	4

Source: Table is from Bruce Hoffman (1998, p. 40). Data source: Schmid *et al.* (1988, pp. 5–6).

political agenda. However, for marginalized groups to be heard, they must often change the scale at which their situation is discussed or decided: groups dominant in a particular state may well have no interest in hearing the complaints of the marginalized. Through acts of terrorism, marginalized groups may change the scope of the political debate, making it a regional or global issue, and so forcing the dominant group in the state to at least talk and maybe even address the situation.

Third, terrorist groups claim, in the words of Hoffman (1998, p. 43), to be performing political altruism. In other words, terrorists believe they are serving or speaking for a group who have been marginalized or oppressed and deserve a better political deal. A more exact understanding of the terrorist would be as a political geographic altruist. The motivation for terrorism is perceived political injustices, but these are inseparable from particular geographic organizations of power relations (see Chapter 1, for a reminder). This is most clear in the case of terrorism motivated by nationalism; the goal is a reorganization of space to create a new independent nation-state. In the case of al-Qaeda, their motivation rests upon the marginalization of Arab influence in the world: specifically, for them, the violence meted out by Israel upon the Palestinians, the exploitation of oil reserves by Western companies, and the presence of US forces across the Arab world. The geographic problem is, broadly speaking, a "colonial" relationship that, it is argued, can be relieved by removing the US presence and eradicating the state of Israel. The killing of Osama bin Laden in May 2011 does not change the fundamental basis for al-Qaeda's existence. The motivation behind terrorism, and hence the possibility for lasting resolution, can only be fully understood through a recognition of the territorial expression of the politics at hand.

Though no single definition of terrorism is possible, the features of the definitions reflect the geography of the causes, and means, of terrorism. Terrorism is an act of geopolitics that is motivated by the spatial manifestation of power, uses geography (in terms of symbolic places and expanding the scope of the conflict) in its tactics, and requires a rearrangement of existing political geographies if it is to be successful or peacefully resolved.

You're a terrorist . . . I'm not

In Chapter 3 we introduced the role of the representations of people, places, and states as an important part of geopolitics. Defining terrorism is also an act of representation that, by restricting the label "terrorist" to a few, creates a wider set of actions and agents that are "non-terrorist." The key question in these acts of representation is the state: some definitions of terrorism are purposeful in emphasizing "non-state" or "sub-national" agents as those who commit terrorism, hence excluding the state as an agent of terrorism (Flint, 2005). Criticizing the omission of consideration of some state actions as terrorism does not imply that every state, throughout history, is a "terrorist." However, restricting terrorism to "sub-national" groups does prevent certain state actions at particular times being designated as acts of violence aimed at instilling fear into the population for political reasons. Such state repression is usually undertaken to establish and maintain control by throttling political opposition. History would, it seems, allow for certain state actions to be seen as the use of violence to create a climate of fear and political compliance.

Adolf Hitler's actions in establishing Nazi Germany and Josef Stalin's political purges are seen as "classic" examples of the state becoming a "police state" to squash any political dissent and opposition. The early example of these states was continued as part of the domestic aspect of the geopolitical codes of states within the Cold War: From the McCarthy trials in the US in the 1950s which brought the power of the state judiciary to bear upon anyone proclaiming a left-wing political agenda and forced people to fear for their careers and reputations, to the secret police forces of the Communist regimes of central and Eastern Europe. The geopolitics of the Cold War constructed domestic "threats" or "enemies within" who were hunted by the state and often tortured and killed, one of the goals being to create a public atmosphere of fear that it was believed would prevent political opposition (see Box 6.6). Contemporary regimes in North Korea, Syria, and many others, some defined as "allies" in the US War on Terror, are guilty of the same actions for the same goals, to varying degrees.

Ahmad's (2000, pp. 94–100) definition of terrorism, purposefully constructed to allow for the inclusion of state actions, has another type of state violence in mind. Ahmad is

Box 6.6 The School of the Americas

During the Cold War the US established the innocuous sounding International Military Education and Training Program (IMET). The Program trained over 500,000 foreign officers and enlisted personnel. The main campus, the School of the Americas, was relocated to Fort Benning, Georgia in 1984. The title of the outfit illustrates that much of the program's regional focus was Central and South America. Defenders of the program claim that it disseminated "American values" through trips to Disneyland and sporting events. However, the product of the school is far from the images of Disney. The school trained soldiers in "low intensity conflict." In other words, not how to fight an invading or hostile army, but how to prevent counter-insurgency in some of the poorest and most polarized countries in the world ruled by undemocratic and brutal military regimes, such as Honduras, Haiti, Paraguay, Uruguay, Chile, Peru, Colombia, Panama, El Salvador, and Guatemala. The School of the Americas includes a "Hall of Fame" displaying portraits of "successful" graduates. "Infamous" would be a more accurate description. To quote Chuck Call of the Washington Office on Latin America, "In El Salvador, 48 of 69 people named in the UN Truth Commission Report as human rights violators, graduates of the school. Half of the people named in a recent report done by NGOs of alleged human rights violators in Columbia, 128 of 247, graduates of the School of the Americas. This is at such a level that you can't ignore it. And what's important about that is that it associates the US military with these abusive forces." Defenders of IMET admit a "few bad apples." Critics of the program argue that the US trains torturers and killers, targeting groups and people who support social reform.

The quotes and information in this box are from a video put out by the American Defense Monitor in 1994 entitled *School of the Americas: At War with Democracy?* The transcript is available at www.cdi.org/adm/804/transcript.html.

In what way does state sponsored torture and oppression fit the definition of terrorism, and in what way can it be argued to be something other than terrorism? How are your answers molded not by what is done but by who (a government agency) is doing it?

referring to the actions of Israel against the Palestinians, and of India and Pakistan in the conflict over Kashmir. In these instances the military wing of the state is using violence in a purposeful and systematic manner to quash nationalist movements that would alter the current boundaries of the state; and in the case of some of the rhetoric and interpretations of the Palestine–Israel conflict, the very existence of the state of Israel. Accusations, inquiries, and revelations still remain over the illegal use of force by the British government against the Irish Republican Army. When the territorial integrity of the state is challenged, the state may go beyond the realms of legality to counter national-separatism. In these situations, violence, diffusing fear through a wider population, and political goals (all common features of definitions of terrorism) are part of the calculations and actions of states. Terrorism? Finally, what of the deliberate and sustained bombing of civilian targets in World War II, as discussed earlier? The goal of these displays of military might was to sap civilian morale and cause surrender. Terrorism?

History of modern terrorism: waves of terrorism and their geography

In a useful, though necessarily simplified exercise, Rapoport (2001) has identified four separate but connected "waves" or periods of modern terrorism. Describing these waves not only offers a brief history of terrorism, but also highlights the changing geography of terrorism (Flint, 2005), a change that has important implications for the contemporary politics of the War on Terror (see Table 6.3).

The goals and arena of the first two waves of terrorism were focused upon one particular geopolitical scale, the nation-state. The first wave occurred between, roughly, the 1880s and the beginning of World War I in 1914 and was motivated by the piecemeal political reforms of the Russian Tsar hoping to preclude more radical and revolutionary change. The goal of the terrorists, loosely defined as "anarchists," was to mobilize the citizens of Russia toward revolution as they feared the population would be placated by the reforms: in other words, the terrorists wanted to change the way that the Russian state was governed. These "anarchist" politics diffused, with limited success, to other parts of Europe. The geography of this first wave was framed by an understanding that the state was the source of political change and so bounded the scope of action. Though the ideology of the terrorists, and the way they conducted terrorism, diffused from Russia into parts of Europe, the geography of the first wave of terrorism was restricted to within state boundaries.

To a lesser degree, the first wave of terrorism also reflected an increase in nationalist politics. The assassination of the Austrian Archduke Franz Ferdinand in Sarajevo by a nationalist sparked World War I, which in turn catalyzed many political and social changes. One of these changes was the explosion of demands for national self-determination, or the desire for people to create and belong to national communities synonymous with independent and sovereign states.

The second wave of terrorism (approximately 1920–60) was dominated by the political geography of ending imperialism, or decolonization, and the establishment of nation-states. Terrorism was, in some cases, deemed a necessary and useful strategy to force colonial powers to leave and, in a related politics, define which social and ethnic groups would play the key roles in defining the new state. Examples of this type of terrorism include the Irgun in Israel angry toward the British government's restrictions on Jewish in-migration, and the Mau Mau in Kenya. The geography of this wave was similar to that of the first; the arena and goal of terrorism was the nation-state, in this case to establish a new one rather than change the politics of existing states. However, more so than the first wave, the impetus toward national self-determination was an agenda that spanned the globe.

The third wave of terrorism (1960s–90s) maintained a nationalist anti-colonial agenda, but with an additional ideological twist. Nationalist groups who saw the project of decolonization and national self-determination as incomplete and unfair resorted to terrorism. Two prominent examples are the Irish Republican Army (IRA) and the Northern Ireland conflict and the Palestine Liberation Organization and its claims for a Palestinian state. The IRA had witnessed the decline of the British Empire across the globe, but called for the process to continue and allow for a united Ireland free of British rule. The PLO had witnessed the establishment of a new nation-state on the territory of Palestine, but it was the state of Israel. In addition to the politics of nationalism was a new component of radicalism, especially in the emergence of terrorist groups in Western Europe and the US motivated by Marxist ideology. For example, the Baader-Meinhof gang in Germany, the Red Brigade in Italy, and the Weather Underground in the US were all motivated by left-wing ideology.

However, the geography of terrorist activity was significantly different from the second wave. In the third wave a greater internationalization of terrorist activity became evident. Terrorist groups were still predominantly based within particular states, and were focused upon change at the scale of the state, but they began to operate and cooperate across state boundaries. The PLO is a good example, using the tactic of hijacking international passenger flights to increase the geographical scope of its activity and generate an international audience for its political message. As airplanes run by British companies sat on the tarmac of foreign airports under the control of Palestinian terrorists and surrounded by non-British security forces, the issue of Palestinian self-determination became more than a problem for Israel and the Arabs. Perhaps the most poignant act was the 1972 Munich Olympic Games when Palestinian terrorists entered the Olympic village, a symbol of international respect and peace, and killed eleven Israeli athletes. Claims of the "whole world watching" were exactly the geographical outcome the terrorists were aiming for: the Palestine–Israel conflict became a matter of international importance and diplomacy.

The second form of internationalization in the third wave was the growing cooperation between terrorist groups based in, and identified with, different states. Training and weapons exchanges became a part of terrorism, and the networks of terrorism became an international rather than national phenomenon. Laqueur (1987) relates the internationalization of terrorism to the Cold War, and the growth in the 1970s of state sponsorship of groups originally defined by their territorial and nationalist demands. Internationalization was perceived by terrorist groups as a means of widening the scope of the conflict and hence increasing the "audience" for their cause. However, it also facilitated state-versus-state conflict. Various governments attempted to gain influence in a particular dispute by supporting different factions of the same cause; such as Syria, Libya, Iraq, and other states funding separate Palestinian groups. The outcome of state sponsorship was to make terrorism "almost respectable," with a sufficient majority of states at the UN preventing any effective international coordination of counter-terrorist actions (Laqueur, 1987, p. 269). The Soviet Union and Libya were significant suppliers of weapons and funds to terrorist groups, but in the 1980s Syria and Iran became increasingly important (1987, p. 295). Today, it is these latter two countries which are the focus of US statements on state-sponsored terrorism and the possible targets of sanctions or military force.

Table 6.3 Geography of waves of terrorism

Period	Terrorist groups	Geography
1880–1914	Anarchists	Within states
1920–60	Nationalists	Within states Decolonization
1960–90	Nationalist Ideological	Internationalization
1990–present	Religious	Transnational "Cosmic"?

The fourth wave of terrorism (1990s–present) portends a much more dramatic geographical change with severe implications for both acts of terrorism and the effectiveness and implications of counter-terrorism. For Rapoport, the fourth wave of terrorism is the period of religious terrorism, though terrorism motivated by nationalism is far from gone. The geography of goals and beliefs of religious terrorists goes beyond international connections; it is a geography that "transcends the state." Perhaps the state as political agent is irrelevant to this form of terrorism.

Christian, Jewish, Muslim, Sikh, and Buddhist religions are all tainted by groups who utilize a fundamentalist view of the belief system to justify acts of terrorism (Juergensmeyer, 2000). In other words, religious terrorism is a contemporary global phenomenon, and not limited to one particular religion, as politically motivated claims against Islam, especially, suggest. Religious terrorists are fighting a "cosmic war"; a war of good against evil in which the adjudicator is God or another form of supreme being, and the terrorists are merely the soldiers conducting God's will (Juergensmeyer, 2000). The battle, in the case of religious terrorism, is for people's souls and not a secular political agenda. The state may be the source of acts deemed "evil" but the state is not the answer; for that, one has to turn to salvation and a different world.

Terrorism motivated by religious fundamentalism is a particularly dangerous form of political violence. It is more likely to invoke terrorist acts that produce a large number of casualties and be less sympathetic to overtures of conflict resolution than the previous waves of terrorism (Juergensmeyer, 2000). Why? To understand this dreary prediction, we have to consider the way the state has dominated both geopolitical practice and analysis throughout the twentieth century. Geopolitical actors have seen the state to be the key structure that either constrains or motivates their actions, but it has also been seen as the key "prize": the geopolitical structure that, if controlled or changed, will reap political benefits. By waging a "cosmic war" religious terrorists have shattered this essential geopolitical assumption of the twentieth century; confounding policy makers and academics in the process.

Religious terrorism, by fighting a "cosmic war" transcends the state as an arena for politics: the goal is to serve God's will and fight "evil"; essentially, the battle is of a spiritual nature and not secular. If that is the case, then victims are "infidels" or "sinners" whose death will, in the minds of the terrorists, please God. With these beliefs, religious terrorists do not need to make the political calculations of secular terrorists in which the number and type of casualties had to be balanced; enough to "shock" but not too many to alienate "sympathizers." For religious terrorists, their actions are part of one sort of Armageddon or another, and not the bloody part of a wider political process, and hence the lack of constraint on the number of casualties.

The second implication of the "cosmic war" thesis is that the state is no longer seen as the key geopolitical arbiter. The state as a structure that could enable terrorists and their sympathizers by providing political concessions, or even conceding defeat, is deemed irrelevant by religious terrorists. The question no longer becomes a matter of harassing politicians to address their concerns, as is usually the goal of nationalist-separatist terrorists. Instead, the belief is that the state is the embodiment of the evil that, following God's will, needs to be destroyed. Again, restraint is not an issue, and the likelihood of large-scale horrifying attacks is increased. For example, Timothy McVeigh did not blow up the Murrah federal building, including the day-care centre, to bring representatives of the US government to the negotiating table; he killed what he saw as agents of evil destroying a "way of life" defined, if loosely, by religious beliefs. For religious terrorists, the state is an actor that needs to be destroyed and not negotiated with. The structure is spiritual and "cosmic," enabling acts of "martyrdom" beyond constraint, if you perceive yourself to be acting on God's will.

But wait a minute. Does religious terrorism really "transcend" the state? There are good reasons to qualify such a claim. Two strands of argument can be made: religious terrorists still use or need states, and the goals of religious terrorism are still related to the state as the key geopolitical structure. The identification of Afghanistan as the "home" or "base" of al-Qaeda immediately after the terrorist attacks of September 11, 2001 is testimony to the relationship between some terrorist groups using religion as their motivation and the need for the protection and sponsorship that can be offered by territorially sovereign states. In the next section we will discuss that relationship between terrorist networks and sovereign states at length. At the moment, it is enough to refer to Bin Laden's past relationships with the governments of Sudan and Afghanistan to see that this particular terrorist group actively sought the haven that territorial sovereignty can provide.

The second question is whether the goals of religiously motivated terrorism transcend the state. For example, interviews with Jewish settlers in the West Bank, and their recourse to scripture for motivation and justification, make for compelling reading (Juergensmeyer, 2000). The belief that the land of Israel was "given" to the Jews by God is clearly part of the conscious that motivates the killing of both secular Jews and Arabs who are deemed to betray or threaten this "return" of Israel to the Jews. But what of the goal? The goal is the establishment of state sovereignty across a particular territory, known as the West Bank or Judaea-Samaria depending on the perspective and agenda. In the British Isles, the conflict in Northern Ireland is usually portrayed as a nationalist struggle, yet Juergensmeyer (2000) emphasizes the religious vitriol between the Protestant unionists and the Catholic republicans. Again, perhaps motivation is being confused with goals. Both sides have agendas regarding the territorial extent of Irish and British sovereignty.

The final issue in discussing whether religious terrorism transcends the states refers to the role of state as arbiter in political disputes. The thesis of "Cosmic war" rests upon the terrorists' perception that God is judging their actions and will provide the subsequent rewards (Juergensmeyer, 2000). But, in some cases, the state has a role to play in evaluating and delivering the terrorists' demands. This is most evident, perhaps, in the case of the United States, where the assassination of doctors performing legal abortions is the extreme manifestation of Christian right lobbying and protest to change the laws of the land and ban abortions. With an increasing number of Senators and representatives in Washington supporting a ban on abortion, it is not inconceivable that access to abortion will be restricted further and even banned. Whether this would be a "victory for terrorism" is a matter of debate. The point is that if such a change in government policy were to be legislated, the goals of terrorists motivated by Christian fundamentalism would have been achieved by the actions of the state.

In summary, terrorism motivated by religious beliefs does appear to be experiencing a surge in activity across the globe and all the major religions. Religious terrorism is creating geography different from those of the previous waves, as the state plays a less central role. Resort to the scale of a "cosmic war" makes religious terrorists less chained to the opportunities and constraints that exist when the state is seen as the key geopolitical structure. This new geography of structure and agency has implications for the severity of terrorist acts and the possibilities for conflict resolution. However, the state is still an essential scale in the calculations of religious terrorists, whether as a strategic territorial haven or the target of political goals. To understand religious terrorism it is useful to think of two separate but closely related geographies: motivation is sought at the "cosmic scale" while goals and actions are still tied to the scale of the state.

Activity

Return to Bin Laden's *fatwa* described in Chapter 2 as a geopolitical code. Define the "spiritual" elements of the code. In what sense do they relate to Juergensmeyer's notion of a "cosmic war"? In what sense is the code focused upon territorial issues that can be interpreted through the political geography of state sovereignty? In what ways, if any, do the spiritual and territorial elements of the code interact?

The importance of religious terrorism in contemporary geopolitics has forced policymakers and academics to rethink the taken-for-granted understanding of geopolitics as inter-state politics. Hence, it requires us to focus upon terrorism and counter-terrorism as involving two, perhaps incongruous, understandings of the world. So, we turn to the meta-geographies of terrorism and counter-terrorism in the next section to show how geopolitics is the interaction between territoriality and the construction of networks.

Meta-geographies of terrorism

We have already discussed the meta-geography of nation-states in Chapter 4. The meta-geography of a network contains two important components: nodes and conduits. The political outcomes of the network are a product of the actions of the people located at different nodes and the way they facilitate flows between nodes. For example, for a terrorist network to function, money, people, weapons, explosives and other equipment, and information must move from node to node. The different nodes in a network will have different functions: training, gathering information, planning, finance, and execution of terrorist acts, for example. Terrorist networks are organized to minimize the amount of contact between nodes so that if one node is identified and engaged by counter-terrorist forces the whole network is not disrupted (Flint, 2003a). Terrorist groups have developed networks in this way over a number of years. For example, the IRA operated different cells of bombers on the British mainland without them knowing of each other's existence. Al-Qaeda is a different model, a network of loosely affiliated movements; perhaps best thought of as an "idea" or common cause than as an "organization" with its implications of centralized control and bureaucratic hierarchy.

An abstract model of a terrorist network requires the definition of particular nodes (commonly referred to as "cells") and the connections (or flows) between them. A terrorist attack requires successful cooperation between cells located across the globe. What are the types of cells in a terrorist network? In what types of places are different types of cells located? How are the cells connected? These questions require the combination of the architecture of networks and the geography of places.

First, the structure of the network must be understood. What may be called "core nodes" are the cells that provide the highest level of planning and purpose of the network. "Peripheral nodes" are the cells that undertake the attacks, the bombers, hijackers, kidnappers, etc. In between are "junction nodes" that translate the plans into action by coordinating funding, training, recruitment, and equipping of the "peripheral nodes." Identifying and destroying the "junction nodes" will maximize the disruption of the network (Hoffman, 2002) because they are the most connected of all the nodes.

To target "junction nodes" they must first be located. The intersection of networks and territory determines particular categories of places that are most suitable for the different types of nodes. Core nodes may be located in territories where state authority is either weak or sympathetic to the terrorists' ideology: in the case of al-Qaeda, southern Afghanistan and northern Pakistan, for example. On the other hand, peripheral nodes must exist and operate in relatively exposed spaces, or those where security is high: airports, borders, secure government and public buildings, etc. The nature of peripheral nodes and the environment in which they must operate makes their "appearance" brief. Also, though destroying a peripheral node will prevent a terrorist attack, the impact on the whole network is limited.

The junction nodes are not only the most connected in the network but also the most exposed; they must have a degree of permanency in relatively exposed spaces. Junction nodes must coordinate the logistics of the network, contacting forgers, arms salesmen, smugglers, financiers, etc. To maintain such contacts requires a relatively stable presence in border zones and cities where security forces may be able to establish surveillance and enforcement presence. In other words, they are the most vulnerable and most important nodes in the network.

Terrorism expert Bruce Hoffman's (2002) identification of a hierarchy of al-Qaeda operatives does not explicitly address the geography of the network, but does point to the differential roles of particular nodes. Hoffman identifies four levels of "operational styles." First is the professional cadre: the well-funded and "most dedicated, committed and professional element" of the group who are tasked with the most important missions. Second are the "trained amateurs" who may well be recruited from other terrorist organizations and have received some training. Their funding is limited and they are charged with "open-ended" missions, e.g. target US commercial aviation rather than a specific target. Third, are the local walk-ins: locally based Islamic radical groups who seek al-Qaeda sponsorship for their own projects. Fourth are the "like-minded" insurgents, guerillas and terrorists: the beneficiaries of Bin Laden's financial "revolutionary philanthropy" and spiritual guidance. The relationship between these groups and al-Qaeda is mutual as they may also offer local logistical support for al-Qaeda operations.

Hoffman's hierarchy of al-Qaeda operatives provides clues to the spatial organization of a terrorist network. Key operatives are trained at particular nodes, and have access to money generated and distributed through another set of nodes. The "trained amateurs" have access to some training nodes but are denied the support of other nodes, especially finance, and so display less connectivity than the "professional cadre." Local logistical support can also be "outsourced" to the "like-minded," preventing the need for all support to come from what could be termed an al-Qaeda network.

What are the implications of such a network organization for counter-terrorism? Hoffman's recommendations reflect an implicit recognition of a hierarchy of nodes in a network. The first recommendation is to target "mid-level leaders" as "Policies aimed at removing these mid-level leaders more effectively disrupt control, communications, and operations up and down the chain of command" (Hoffman, 2002, p. 21). In other words, these leaders staff important nodes in the network, that facilitate the combination of plans and resources that make a terrorist attack happen. In network terms, Hoffman is proposing the targeting of a junction node that, once gone, negates the efficacy of all other nodes.

Hoffman's second recommendation is to "De-legitimize – do not just arrest or kill – the top leaders of terrorist groups" (2002, p. 22). The argument is that leaders do more than coordinate a network, they give ideological purpose to its existence. By portraying the leader as corrupt or hypocritical the ideological glue binding the network together may loosen. In addition to killing bin Laden, the US repeatedly showed images of him "unkempt" and questioned the nature of his arranged marriages. He was still being delegitimized after his death.

The third recommendation is to "[f]ocus on disrupting support networks and trafficking activities" (2002, p. 22). The terrorist requires a network of support; if these support-ing connections are disrupted (and they may be easier to identify and arrest) then the final node of the network is starved of what it needs. Hoffman's fourth recommendation is to "Establish a dedicated counter-intelligence center specifically to engage terrorist reconnaissance" (2002, p. 23). Reconnaissance may be either the sole task of a particular node or one of the tasks of the ultimate perpetrators, but it requires a degree of visibility at what is likely to be a well-policed location. These last two counter-terrorism recommendations recognize that certain nodes are more vulnerable than others, and make for more profitable counter-terrorism.

It is not just the function or type of the node that is crucial; it is also the geographic context in which it operates. For example, Hoffman's (2002) recognition of reconnaissance activities is given further import because of the need for a terrorist to spend time in a

Figure 6.6 War on Terror.

well-policed location. The coordinating role of mid-level leaders may require a certain fixity and visibility at a particular location that abets counter-terrorism. On the flip side, the ideological function of leaders allows them to retreat to geographical areas that are hard to police – the tribal areas of Pakistan, for example. Finally, the merging of terrorist networks with other criminal activities, such as smuggling, requires terrorist networks to operate in border zones that may facilitate counter-terrorism. The geography of the terrorist network is laid over maps of policed territories, and the variation in the level of policing across space. Terrorists try to locate nodes with this geography of policing in mind. Counter-terrorist agencies try to identify where nodes are forced to become the most visible.

Terrorists have created a meta-geography of the terrorist network in order to fight power organized in a different and established meta-geography, territorial sovereign states (Flint, 2003a; 2003b). In the first three waves of terrorism, networks were mainly organized within a particular state, hence the jurisdiction of counter-terrorist forces overlay the spatial extent of the network. However, during the third wave of terrorism this geographical relationship began to change, as training, especially, was conducted in foreign countries. Cooperation between states (such as that between France and Spain to counter ETA) was relatively easy as they were neighbors with a common interest against the terrorist group. The internationalization of the PLO was a different matter, operating in either states or territories that did not facilitate cooperation between states. The current War on Terror has made the situation much harder for states. The goals of al-Qaeda are hard to discern and the geography of the network has been difficult to identify. Even when it appeared that operating cells within the US were identified, some of these allegations have not stood up to judicial scrutiny.

Incongruous geographies?

The larger meta-geographic point is that, in order to counter a terrorist network, the United States has had to conquer sovereign territory (Flint, 2003b); or the geopolitics of state territoriality and networks clash. The methods of terrorism and counter-terrorism construct very different, even incongruous, geographies that have implications for the success of counter-terrorism. States must challenge networks by controlling sovereign territory. More than just being inefficient, this may actually be a counter-productive counter-terrorism as it increases the presence of US forces in other countries. As a result, bin Laden's *fatwa* becomes prophetic.

The primary purpose of the invasion of Afghanistan and the overthrow of the Taliban regime was the disruption of al-Qaeda bases: a sovereign state was invaded to destroy the nodes of a network. The operation has been only partially successful as some territory has remained beyond the control of the US and allied security forces, namely eastern parts of Afghanistan and northern parts of Pakistan. Moreover, the US's territorially based response to the attacks of September 11th, 2001 have reinforced the rhetoric of al-Qaeda that views the United States as conducting a global "crusade" against Muslims. The strategy of controlling territory to combat a network not only has reinforced the perceptions of al-Qaeda sympathizers that the US is on a global mission, but also has relocated US troops and made them potential targets (refer back to al-Qaeda's geopolitical code on pp. 62–63). Figure 6.7 shows the extension of US bases into Central Asia immediately after the terrorist attacks of September 11, 2001.

The same strategy was used to justify the 2003 invasion of Iraq, though subsequently President George Bush's administration admitted there were no connections between al-Qaeda and Saddam Hussein. Saddam Hussein was portrayed as a ruler who was using the territorial sovereignty of Iraq to facilitate the maintenance of the al-Qaeda network. Justification for the war rested upon the need to invade the sovereign territory of Iraq to disrupt a network that had some, poorly defined connection with Iraq, and might use those connections to conduct further attacks within the sovereign territory of the United States. Simply put, the US argued that disrupting a terrorist network required the military invasion and occupation of sovereign territory.

However, it is also evident that the War on Terror is using fewer territorial tactics to counter the terrorist network al-Qaeda. First, cooperation with other countries has met with some success as arrests of alleged terrorists have been made in Pakistan and Indonesia, for example. Less conventionally, and with greater geopolitical implications, is the use of aterritorial weaponry to target alleged terrorists in other sovereign spaces. However, the operation to kill Bin Laden has exposed the geopolitical complexity of such an approach. Supporters of Bin Laden and some other groups in Pakistan have criticized the US for breaching Pakistan's territorial sovereignty. The US claims that there was an agreement in place with the government to allow such an operation (see Box 6.7). The weaponry of drones, both to observe but also to attack, has allowed the United States to be "present" in accessible areas without a physical military capability "on the ground." Though initiated by President George W. Bush, the rate of drone attacks, especially in Pakistan, increased during President Barack Obama's administration. Drones are operated from afar; often the control room is in the United States, and relies upon the judgment

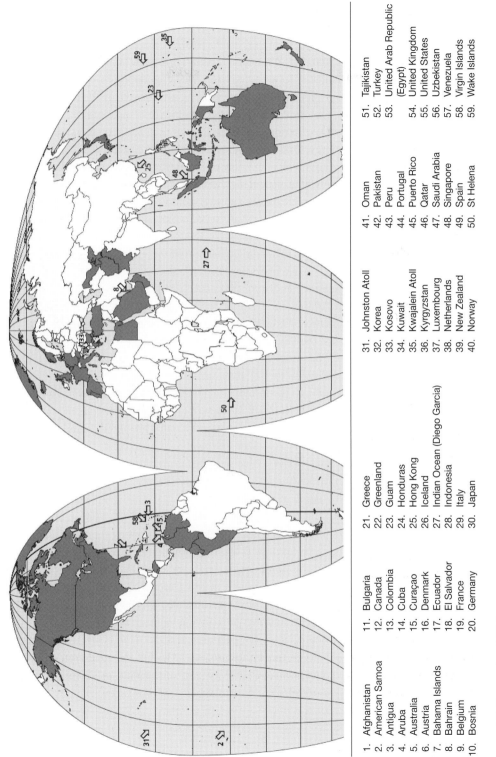

1. Afghanistan
2. American Samoa
3. Antigua
4. Aruba
5. Australia
6. Austria
7. Bahama Islands
8. Bahrain
9. Belgium
10. Bosnia

11. Bulgaria
12. Canada
13. Colombia
14. Cuba
15. Curaçao
16. Denmark
17. Ecuador
18. El Salvador
19. France
20. Germany

21. Greece
22. Greenland
23. Guam
24. Honduras
25. Hong Kong
26. Iceland
27. Indian Ocean (Diego Garcia)
28. Indonesia
29. Italy
30. Japan

31. Johnston Atoll
32. Korea
33. Kosovo
34. Kuwait
35. Kwajalein Atoll
36. Kyrgyzstan
37. Luxembourg
38. Netherlands
39. New Zealand
40. Norway

41. Oman
42. Pakistan
43. Peru
44. Portugal
45. Puerto Rico
46. Qatar
47. Saudi Arabia
48. Singapore
49. Spain
50. St Helena

51. Tajikistan
52. Turkey
53. United Arab Republic (Egypt)
54. United Kingdom
55. United States
56. Uzbekistan
57. Venezuela
58. Virgin Islands
59. Wake Islands

Figure 6.7 Geography of US bases.

Box 6.7 Sovereignty and counter-terrorism

The stunning news on May 1st, 2011 that Osama bin Laden had been killed in a house in Abbottabad, Pakistan soon turned to a series of geopolitical questions that centered upon territory and sovereignty. The government of Pakistan was embarrassed by the fact that Bin Laden had been "hiding in broad daylight," though in reality behind the walls of a compound, in a town that was also the site of the Pakistani Military Academy. Questions were asked about the extent of Bin Laden's "support network," or who in Pakistan knew of his presence. Allegations about the complicity of the Pakistani state, especially the Pakistan intelligence service (Inter-services Intelligence or ISI), were made. The suggestion being that Pakistan, at best, was unable to effectively police its territory or, at worst, was providing territorial sanctuary to the world's most wanted terrorist.

Protest in Pakistan revolved around the question of the violation of sovereignty that the US had enacted in an operation that involved a helicopter attack by Special Forces. For a sense of Pakistani discontent, imagine the reverse situation: a Pakistani helicopter landing in a US or European suburb and commandoes conducting a killing with no consideration that they may be brought before the law. To reduce popular protest the Pakistani government made some claims that it was not informed about the operation, though former President Musharraf stated that he had made an agreement with the US that such an attack was permissible if and when Bin Laden's location was identified. Greater questions regarding sovereignty were raised when some reports claimed that a supporting US force was on hand in case the operational team needed to "fight its way out," presumably including a potential confrontation with the Pakistani military.

The timing of the killing of Bin Laden in Pakistan coincided with international criticism (notably from Russia) regarding NATO missile attacks upon the home of Libyan leader Colonel Muammar Gaddafi as part of the West's support for those opposed to his regime in what had developed into a civil war. The criticism claimed that NATO had violated international law by deliberately targeting for assassination the leader of a state. Such an attempt is a gross violation of sovereignty. NATO denied the charges, saying the home was a "military target." The killing of Bin Laden and what appeared to be an attempt to kill Colonel Gaddafi illustrate the relative ease with which the sovereignty of some states can be violated, and how it is largely accepted as a contemporary geopolitical tactic or reality.

As we shall see in the next chapter, the geopolitical code of the US is particularly focused upon "global reach" or the ability to operate within the sovereign spaces of others. This tendency has been intensified as part of the War on Terror.

of an observer seeing remote images on a screen. The use of drone or remote weapons suggests that warfare has reached a new threshold in which traditional territorial constraints are increasingly irrelevant. This is most obvious in the case of cyber-warfare, the last form of network geopolitics we will discuss.

Geopolitics of Netwar and cyber-warfare

The pervasiveness of networks, surveillance, and various forms of knowledge has promoted a new form of warfare and a geopolitics of security that is framed around the capabilities and vulnerabilities of networks. The term Netwar (Arquilla and Ronfeldt, 2001) has been adopted to refer broadly to the role of networks in conflict, and includes the actions of social movements, terrorist organizations, and computer systems. Cyber-warfare has a more specific focus: the use of computers to attack other computers and networks through electronic, rather than physical, means (Billo and Chang, 2004). The increasing relevance of cyber-warfare is a result of the growth of "information" in the operation of society, economics, and the conduct of warfare. Information stored, organized, and analyzed by computers is necessary infrastructure in the contemporary world, and if its usage can be disrupted then it may be interpreted as an attack on a country or business.

National security institutions, private companies, and a variety of "experts" have readily identified threats and actual incidents of cyber-warfare (Clarke, 2010). Numerous examples exist: in a bout of tit-for-tat hacking of Pakistani websites by an Indian group and vice versa in November and December 2010; and in May 2008 a hacking incident in a US military installation in the Middle East which led to the diffusion of computer code that provided a "beachhead" for the continued transfer of data from US military computers (Lynn, 2010). William J. Lynn III, US Deputy Secretary of Defense, boldly claimed that 100 foreign intelligence agencies are actively attempting to hack into the US's military and intelligence computers (2010). The form of conflict also, allegedly, involves states versus private companies. Especially there have been allegations of attacks emanating from China on Google and other companies. The Chinese government has denied responsibility.

Transport and energy infrastructure is seen as being especially vulnerable to cyber-attack, with scenarios of blackouts as the electrical grid is hacked. These scenarios were given some credence by the Stuxnet attacks on the Iranian nuclear facility at Natanz in September 2010, allegedly severely disrupting Iran's nuclear program. Stuxnet is a form of malware that enters industrial systems, can transfer information out of the system (or "spy"), and disrupt its commands. Iran, and the facilities involved in uranium enrichment, have been the focus of Stuxnet attacks. Kapersky Labs believe that the attacks could only have been done by, or with the assistance of, a state; with suspicion falling upon Israel, given its fears of a nuclear-armed Iran (Maclean, 2010).

Another form of cyber-warfare has been the influence of Wikileaks, or the ready dissemination of information deemed secret and private by states and businesses that reveals allegedly criminal and immoral acts, or just episodes of diplomatic ineptitude and ineffectiveness. The ability of Wikileaks to obtain and spread information has had an impact upon the sense of privacy that states have held in their conduct of foreign policy. Though states have always been the victim of spies, information gained through espionage has rarely been disseminated through the public realm. Now states are aware that the nature of their activity can become public, and diplomacy may have to be adjusted accordingly.

The contemporary geopolitics of surveillance is one of ability and vulnerability. The ability to track through drones and computer surveillance enhances the geographic reach of states. On the flip side, ever-increasing reliance upon computer and information networks has exposed states to attack through the infiltration of viruses, malware, and spyware. A new form of conflict has emerged involving hacking. Much of this activity is undoubtedly state sponsored, but the geography is such that servers in different countries may be used and the identity of protagonist states hidden. As the ability of states to manage and control their territorially based populations is increasingly dependent upon computer networks, a hacker using a global network of servers makes disruption increasingly likely and hard to pin-point geographically.

Summary and segue

Our discussion of the geopolitics of transnational social movements, terrorism and counter-terrorism, and cyber-warfare has illuminated some key points. The main point of emphasis is that geopolitics involves the dynamic interaction between territories, most notably states, and networks. Perhaps increasingly, if the intensification of globalization continues, the opportunities for political action and the threats posed by network activity will come to the fore. States have traditionally identified their allies and enemies, or opportunities and threats, in terms of other states. But this calculation may well become increasingly irrelevant or incomplete as various forms of networks come to the fore. However, states are not about to disappear entirely. Geopolitics will need to develop ways of critically assessing how and why territory and networks interact to form new political circumstances. Such a re-evaluation of security requires an understanding of the context or structure within which change is occurring. In the next chapter we investigate one way to think of geopolitical structure.

Having read this chapter you will be able to:

- Consider the geopolitics of globalization
- Understand the activity of social movements as a form of geopolitics
- Interpret peace movements as geographically situated actors
- Identify the geography of contemporary terrorism
- Identify the geography of contemporary counter-terrorism
- Consider the geographic mismatch of terrorism and counter-terrorism
- Consider the geopolitical implications of cyber-warfare

Further reading

Adolf, A. (2009) *Peace: A World History*, Cambridge, UK and Malden, MA: Polity Press.

A thought-provoking definition of peace, linking a variety of scales and processes and a compelling analysis of how peace activism has proven effective across the course of human history.

Flint, C. (2005) "Dynamic Metageographies of Terrorism: The Spatial Challenges of Religious Terrorism and the 'War on Terrorism'," in C. Flint (ed.), *The Geography of War and Peace*, Oxford: Oxford University Press, pp. 198–216.

A discussion of terrorism and the War on Terror emphasizing the interaction between network and nation-state metageographies.

Herb, G. H. (2005) "The Geography of Peace Movements," in C. Flint (ed.), *The Geography of War and Peace*, Oxford and New York: Oxford University Press, pp. 347–68.

An accessible analysis of the geopolitical contexts that have led to the formation of peace movements across history, and the changing geographic strategies they have adopted.

Hoffman, B. (1998) *Inside Terrorism*, New York: Columbia University Press.

An excellent and accessible introduction to the study of terrorism.

Juergensmeyer, M. (2000) *Terror in the Mind of God*, Berkeley: University of California Press.

A thought-provoking analysis of the motivations and implications of terrorism conducted by religious fundamentalists in all the major religions.

7

GLOBAL GEOPOLITICAL STRUCTURE: FRAMING AGENCY

In this chapter we will:

▦ Introduce a geopolitical model to provide an understanding of the global geopolitical structure

▦ Discuss the different components of this model

▦ Interrogate the validity of the model

▦ Note how the model is both similar to and different from "classic" geopolitical frameworks

▦ Emphasize how we can use the model to provide a structure or context to understand geopolitical agency

---

Let us take some time to consider how we began this book and our exploration of geopolitics. In the prologue we learned about the traditional practices of classic geopolitics and its claim to be able to paint neutral and complete pictures of "how the world works": what drives historical changes, what causes countries to fight, what determines whether a country will become a great power or not. In Chapter 1 we introduced a framework to analyze geopolitics objectively through the use of geographic concepts and a consideration of structure and agency. The bulk of this book has focused on agency by introducing the term geopolitical codes, and showing how it is related to nationalism, and the geography of territory and networks.

In this chapter we begin to discuss geopolitical structure, or the context within which geopolitical agency takes place. We do so by discussing George Modelski's model of cycles of world leadership. One of the benefits of this model is an understanding of global politics that is based on empirical observation. In contrast, the classical geopoliticians of the nineteenth and early twentieth century invoked a "God's eye view of the world," providing simple histories or theories that, they claimed, not only explained what has happened in the past, but suggested particular policies to inform the actions of their own country in a global competition with others (Parker, 1985). In other words, geopoliticians made dubious claims of historical and theoretical "objectivity" to support their own biased view of how their own country should compete in the world.

Box 7.1 Confucius Institutes

"Chinese Move to Eclipse US Appeal in South Asia," claims the *New York Times* headline (November 11, 2004, A1) for an article by Jane Perlez talking of university students in Thailand choosing to learn Mandarin over English as a result of perceptions that China will be the dominant influence in the region soon. The Chinese have funded and built many language and cultural centers "as part of China's expanding presence across Southeast Asia and the Pacific, where Beijing is making a big push to market itself and its language, similar to the way the United States promoted its culture and values during the cold war." These language centers include the Confucius Institutes and are directly funded by the Chinese government.

In a related report by the Stimson Center in Washington, DC the establishment of Confucius Institutes by China is contrasted to the declining investment by the US in means to disseminate its cultural power, such as Voice of America radio. The report finds that most of the hundred or so Confucius Institutes are located in North America and Europe, and of the seventeen in Southeast Asia fifteen are in Thailand. The Stimson report is called "Confucius Institutes and Chinese Softpower in Southeast Asia" and can be accessed at www.stimson.org/spotlight/confucius-institutes-and-chinese-soft-power-in-southeast-asia-1/. Posted April 3, 2008, accessed May 5, 2011.

Such a view of geopolitics is no longer in vogue. Any claim to be able to "see" a pattern of global politics is immediately challenged as being limited and biased; rightly so – because it is situated knowledge. Instead, attention is drawn to how geopolitical agents make strategic choices, and how these are complicated by competing goals and changing circumstances. In other words, increasing attention is given to agency over structure. However, decisions are not made within a social and political vacuum. As discussed in Chapter 1, agents are both enabled and constrained by structures. Countries make geopolitical choices, to go to war for example, while considering the wider geopolitical context. For example, China's increasing political and economic power has led to greater influence within East and Southeast Asia. The Chinese government, on the other hand, is being very careful not to provoke the dominant world power, the United States (see Box 7.1).

As this example suggests, another benefit of Modelski's model is its ability to aid our interpretation of the role of the United States in the world. In other words, we can understand the US as a particular type of geopolitical agent, a world leader. Furthermore, the model suggests that the world leader plays a key role in creating a global geopolitical structure and that the way that structure changes over time is a way to interpret, or place into context, the geopolitical codes of the US, other states, and non-state actors. Geopolitical decisions are made with an eye toward the global geopolitical context, and especially the ability of a dominant power to set the agenda.

In this chapter, we will introduce Modelski's model of geopolitics to define a global geopolitical structure. We will see that this structure is dynamic and use it to discuss how the global geopolitical context frames the actions of different countries. Though the chapter ends with a guide to allow for critique of the model, it may be useful to provide some cautionary notes here. Modelski's model of geopolitics is *not* capable of *predicting* events. It is a historical model that interprets a wealth of historic data in a simplified framework. In other words, it is a *descriptive* model. Also, Modelski's model is useful, but only within certain parameters. His view of geopolitics is limited to conflicts between the major powers; smaller countries and geopolitical actors that are not countries are not included in his model. However, the model is useful for introducing the idea of a geopolitical structure and offering a context for current geopolitical events. We will discuss the pros and cons of the model in greater depth at the end of the chapter.

Defining a global geopolitical structure: using and interrogating Modelski's model of world leadership

Mahan and Mackinder, as well as Ratzel and Kjellen in Germany, exemplified the state-centric perspective of geopolitics, and geographical determinism (as discussed in the Prologue). From their perspective, geographic size and location and the internal make-up of a country determined power. Subsequent, and purportedly more scientific, calculations of power have rested upon the economic, military, and demographic elements of a particular country. To understand state power and global geopolitical context, however, these ingredients must be related to the ability of a state to define the global geopolitical agenda. In other words, the Gramscian notion of power within a country that we introduced in Chapter 1 has relevance for global geopolitics. Following Gramsci, we would expect the most powerful countries to wield (or at least attempt to wield) an ideological power over the other countries: the most powerful country would try to set a political agenda that the rest of the world would, more or less, follow. Two theories have been particularly influential in the discussion of this type of global agenda setting: Wallerstein's concept of hegemony (see Box 7.2) and, the one we will engage, Modelski's (1987) concept of world leadership.

Modelski's model of world leadership is a historically based theory, founded upon his interest in naval history. Power, for Modelski, is a function of global reach – the ability to influence events across the world. In history, such power has required control of the oceans. Hence, for Modelski, world power rests upon the ability of one country to concentrate ocean-going capacity under its own control. Ocean-going capacity is measured by the combined tonnage of a country's military and merchant navies. In this sense, Modelski echoes Mahan's insistence on the important role of sea-power. However, most significantly, for our understanding of the contemporary world, world leadership is not defined solely by this material measure of power. Indeed, it is important to reflect upon the name Modelski gives to dominant and powerful countries – they are identified as world leaders, not hegemonic or super powers. Remember, a crucial component of geopolitics is representation. Modelski portrays the world's most powerful country as a "leader,"

Box 7.2 Wallerstein's world-system theory

The sociologist Immanuel Wallerstein (1979; 1984) profoundly challenged modern social science through his concept of the historical social system. His argument was that society should not be equated with a particular country, but rather at a larger scale of the social system. According to Wallerstein, since approximately 1450 the social system has been the capitalist world-economy. Within this theory, primary geopolitical powers are called hegemonies or hegemonic powers. Since the twentieth century, the United States has acted as hegemonic power. The basis for hegemony is economic strength that translates into a dominant influence in global trade and finance. Maintenance of the capitalist world-economy in a form that benefits the hegemonic power requires, at times, military force. Hegemony is seen as an economic process for selfish goals, and not the global political benevolence of Modelski's world leadership. Similar to Modelski's model, the hegemonic power emerges from a period of global conflict, but Wallerstein is adamant that the United States is currently experiencing a relative decline in its global dominance. One other important difference is that in Modelski's model there is always a world leader, though its strength is cyclical. For Wallerstein, periods of hegemony are rare. So, if the US's hegemony does decline, according to Modelski a new leader should emerge after a period of war. Wallerstein's model suggests that other political scenarios, without one dominant state, may emerge.

Table 7.1 Cycles of world leadership

World leader	Century	Global war	Challenger	Coalition partners
Portugal	1500s	1494–1516	Spain	Netherlands
Netherlands	1600s	1580–1609	France	England
Great Britain	1700s	1688–1713	France	Russia
Great Britain	1800s	1792–1815	Germany	US plus allies
United States	1900s	1914–45	Soviet Union/ al-Qaeda	NATO/Coalition of willing

Source: George Modelski (1987)

Activity

Is it more accurate to think of the United States from the early twentieth century to today as a hegemonic power or as world leader? To answer the question, think about the relative weight given to economics versus politics and self-interest and political duty in the different models. Perhaps you may be able to find examples of both. Is the global "responsibility" of the US as either world leader or hegemonic power solely a matter of rhetoric, or can you also point to particular actions?

implying willing followers, rather than a hegemonic or super power with its allusions to dominance and force.

Obviously, Modelski's definition of power is of the ilk that is strongly criticized by feminists (see Chapter 1). Power, in the model, is about strength and dominance; it is about the ability to exercise military force across the globe. This is another way in which Modelski follows the "classic" geopoliticians. This notion of power leads to an uncritical belief that the militarization of foreign policy is inevitable and beneficial. It also ignores gender relations within states and global economic inequities. In other words, Modelski's notion of power is uni-dimensional. We may agree that a feminist critique of Modelski's power index is valid and yet still find value in the model. In fact, we have seen in the previous chapters that geopolitics is represented in certain gender specific ways for the power relations Modelski identifies to be sustained. In other words, by bringing a feminist critique to bear upon Modelski we can get more out of the model than was originally intended by its author.

A world leader is a country that is able to offer the world an "innovation" to provide geopolitical order and security. By innovation Modelski means a bundle of institutions, ideas, and practices that establish the geopolitical agenda for the world. The power of the world leader rests in its ability to define a "big idea" for how countries should exist and interact with each other; an idea that it is able to put into practice through its material power or naval capabilities. The power of the world leader rests in its agenda-setting capacity and its ability to enforce it.

Modelski's model of world leadership is dynamic. The strength of the world leader rises and falls. Over the course of centuries, the mantle has passed from one country to another in a sequence of cycles of world leadership (Table 7.1). Each cycle lasts approximately one hundred years and is made up of four roughly equal phases of about twenty-five years:

Phase of global war: The ability, or perceived right, to act as world leader is decided through a period of global war. The declining world leader is challenged by countries believing they should inherit the mantle. Coalitions are constructed, and over the twenty-five-year period, that may include a number of different wars and conflicts, one country emerges as having both the material capacity and the ideological message to impose global order.

Phase of world power: Once victory has been achieved the geopolitical project of the new world leader is enacted. New institutions are established to apply and enforce the new agenda. On the whole, the new agenda is welcomed and followed.

Phase of delegitimation: At the outset of the establishment of a new period of world leadership, the imposed "order" is, overall, welcomed. But over time dissent grows. The benevolence of the world leader can be questioned; its actions seen increasingly as self-serving. Alternative agendas are given greater weight. The challenge to the world leader has begun, but the world leader is still relatively strong.

Phase of deconcentration: The challenges beginning in the previous phase become stronger. The world leader expends its material and ideological capacity in reacting to these challenges, making it weaker and more vulnerable to more attacks, in a spiral of challenge and reaction that leads to the phase of global war. Challenges are more frequently, but not exclusively, violent and organized campaigns. The world leader is

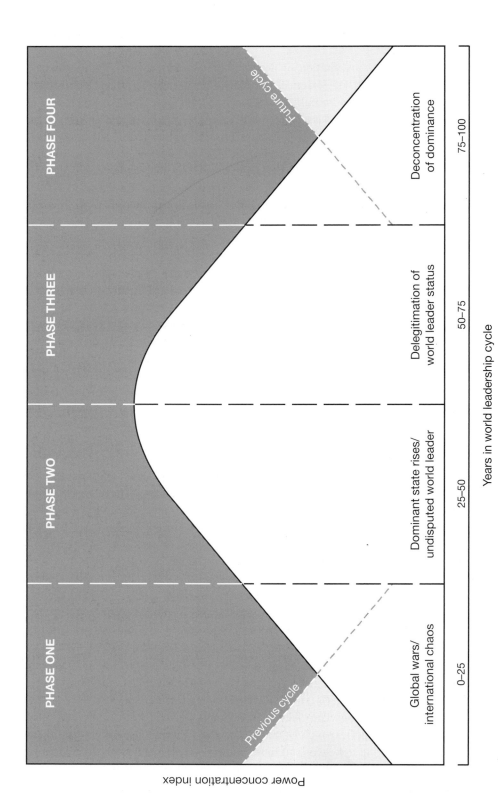

Figure 7.1 Modelski's world leadership cycle.

called upon to react militarily, exhausting its material base of power and highlighting contradictions between its actions and its rhetoric. In combination, its legitimacy is increasingly questioned, and challenge intensifies.

As a concrete example, the War on Terror has been represented by the United States as a mission in the name of "peace" and "humanity." As part of that military action, atrocities have come to light that add fuel to the fire of those opposed to the US presence in Iraq and Afghanistan. A prominent event was the abuse in Abu Ghraib prison that provided a sharp contrast between the representation of the occupation of Iraq as a "civilizing mission" of world leadership and actual events that challenge the leader's authority (see Box 7.4).

Using the ideal conceptual framework we have discussed, Modelski paints a particular picture of history – one defined by the cycles of world leadership. The role of representation in his model is most important. In a cold use of language, global wars are defined as a "systemic decision" – they are instrumental in deciding who will be the next world leader (see Box 7.3). For Modelski, a leader is seen as acting benevolently – carrying the burden of maintaining global security for the benefit of all rather than acting for narrow national self-interest. The order defined by the "innovation" is portrayed as neutral; it is seen as being obviously good for all, rather than benefiting some countries or groups over others. Perhaps most significant is the pattern of history Modelski identifies from the application of his model. Great Britain was able to have two consecutive cycles of world leadership. The geopolitics of the model is clear: if the Brits had two shots then there is nothing stopping the United States doing the same thing; the twenty-first century can be an American century too!

Box 7.3 World Wars I and II in historical context

Both Modelski and Wallerstein view the two World Wars as twin episodes in one conflict, the one that decided who would succeed Great Britain as world leader/hegemonic power. Modelski is also guilty of representing these two (or is it one?) conflicts in cold language. Together, they are identified as a "systemic decision" of world leadership succession – a very instrumental way to view the deaths of millions of soldiers and citizens across the globe.

Within the phase of global war, the emerging world leader has a "good war," in the sense that it avoids much of the physical destruction of its homeland suffered by other fighting countries. Hence, its relative economic power increases dramatically. In the case of World War II, as the factories of Germany, Japan, and Great Britain were being flattened by aerial bombing, those in the United States were expanding their capacity. The emerging leader also enters the conflict relatively late – using its relative power to dictate the terms of peace to its liking. For further reading see Peter Taylor's use of Wallerstein's framework to analyze how Great Britain faced opportunities and constraints in creating its post-World War II foreign policy in his book *Britain and the Cold War*.

The geopolitics of the rise and fall of world leaders: the context of contemporary geopolitics?

Modelski's model helps us to interpret the major contemporary global geopolitical issue: the attempt by the United States to maintain its preeminent power status in the face of challenges to its leadership. To do this we can consider the dynamics of two separate but related concerns. First, is there a country willing and able to act (Modelski may well say "serve") as world leader? In other words, is there an availability order, the possibility of one country, the world leader, to offer and enforce a geopolitical innovation? Second, does the rest of the world, or at least a significant majority, want that order? In other words, is there a preference for the world leader's imposed order, or would countries rather face the "chaos" or "insecurity" of competing agendas? Note the role of representation here again – as "insecurity" and "security" is often based upon the degree of acceptance of the world leader's agenda.

For each of the four phases of a cycle, we can compare the balance of preference and availability of order (Figure 7.2). In a period of global war, no one country is strong enough, relative to others, to establish a global geopolitical order. After the emergence of a world leader, there is a desire for order and the world leader's agenda is followed, more or less. By the next phase, delegitimation, the order being provided by the world leader is beginning to be questioned. However, the world leader still retains its relative power advantage, and hence challenges to the world leader rest, on the whole, in the realm of diplomatic and verbal protest; though some sporadic military resistance may be witnessed. During the deconcentration phase of the cycle, not only has dissent toward the world leader's order heightened, but the world leader's ability to enforce its agenda has declined too. In this phase, there is an increased challenge to the world leader, not only in terms of diplomatic and political agendas, but also in the form of organized military challenges.

Imperial Overstretch

Global opinion is only one factor in explaining the process of the decline of world leadership. Emphasis has also been placed upon the relationship between the demands placed upon the cost of the world leader's military and its economic strength, or the ability to

Modelski phase	Preference for world order	Availability of world order
Global war	High	Low
World power	High	High
Delegitimation	Low	High
Deconcentration	Low	Low

Figure 7.2 Preference and availability of world leadership.

Great Britain's military expenditures

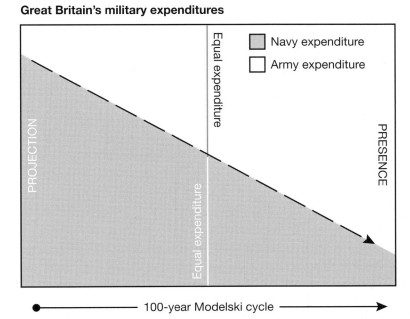

Figure 7.3 Imperial overstretch.

pay. During the world leadership phase of the cycle, where the new global agenda is mostly accepted, enforcement can be attained by a global naval capacity – the strategy of gunboat diplomacy whereby the mere presence of the world leader's navy is enough to keep potentially dissenting countries in line. Such a strategy is relatively cheap, as the very costly undertaking of protracted military conflict is largely avoided. However, as the cycle progresses, and challenges to the world leader's authority increase in frequency and intensity, then the world leader is drawn increasingly into conflicts on land. Associated rising costs further drain the world leader's power and invite more challenges. In addition, the ghastliness of warfare provokes specific incidents which are used by opponents to challenge the moral authority of the world leader (see Box 7.4). In other words, the resort to increased land conflict is costly in both economic and ideological terms.

Evidence of imperial overstretch?

The idea of imperial overstretch is a useful framework to evaluate the policies of the US, as world leader. A continuous topic of debate concerning the war on Iraq was the number of US troops needed for the initial invasion and the subsequent "post-invasion" goal of forming a democratic Iraq. Secretary of Defense Donald Rumsfeld was continually criticized for not deploying enough troops. In the US military organization, any large military operation, such as the invasions of Afghanistan and Iraq, relies upon the deployment of the National Guard, a military reserve of citizens. The deployment of these soldiers (men and women) has an impact upon families, communities, and businesses as

Box 7.4 Abu Ghraib and the consequences for world leadership

In the spring of 2004 a series of photographs hit the global media that shattered the US's attempt to portray its occupation of Iraq as a humanitarian mission of the world leader intent on promoting human rights. Soldiers had taken pictures of practices in Abu Ghraib prison in Iraq of inmates being subjected to demeaning and painful acts, tantamount to torture, designed to break their resolve prior to interrogation. The images included those of a terrified inmate warding off a prison guard's attack dog, an inmate forced to kneel as if performing oral sex on another, laughing soldiers standing by a pyramid of naked detainees, inmates being led around on leashes by prison guards, and a hooded man posing as if on a crucifix with electric wires attached to his hands and penis.

In his investigation of the US abuses, Major General Antonio Taguba found practices which included:

> Breaking chemical lights and pouring the phosphoric liquid on detainees; pouring cold water on naked detainees; beating detainees with a broom handle and a chair; threatening male detainees with rape . . . sodomizing a detainee with a chemical light and perhaps a broom stick, and using military working dogs to frighten and intimidate detainees with threats of attacks, and in one instance actually biting a detainee.

Ideologically, Abu Ghraib was a disaster for the US's portrayal of itself as world leader, and will have lasting impacts. The photographs alienated politicians in foreign countries who supported the US's mission of creating democracies through military presence, or put them into a position where they could no longer support US actions because of negative public opinion. Obviously, the images inflamed those already opposed to the US's role in Islamic countries, and were used to justify their existing rhetoric of Americans as "infidels."

The abuse in Abu Ghraib, along with allegations of torture at Camp X-Ray, Guantanomo Bay, was not inevitable or predetermined. However, these acts were a product of the world leader's self-imposed policing mission, and the dissemination of the images of torture undermined the ideological authority underlying its position. In other words, the actions of the world leader contradicted its rhetoric.

In May 2011, just a few days after Osama bin Laden had been killed, two questions came into focus: should an apparently gruesome picture of Bin Laden's body (with bullet shots to the face) be released, and was the mission designed to kill him with no attempt to capture? The first question was partially provoked by an atmosphere of conspiracy theory within the US and a tendency to distrust the government's claim that the terrorist had been killed. President Obama resisted releasing the photo because of the negative reaction it would likely induce in parts of the world that supported al-Qaeda. The second question was still being debated at the time of writing, but was being raised (by the UN special rapporteur on extrajudicial, summary, or arbitrary executions among others) to delegitimize US actions as a form of "summary justice." The prosecution of the War on Terror continues to raise questions of the morality of the actions of the world leader.

Figure 7.4 Camp Delta.

mothers, fathers, neighbors, employees, etc. are sent overseas. The impact of fighting has long-term effects too, as recruitment into the military and the reserve may decline.

The dynamics of US troop deployment may also shed light upon the dynamics of world leadership (see Table 7.2 to see both stability and change in the geography of US global troop deployments). The 2009 Army Posture Statement, an official publication of the US Army, begins by noting that the Army has been "stressed by seven years of war" (Geren and Casey, 2009, p. 1) The report goes on to say: "The Army is out of balance. The current demand for our forces in Iraq and Afghanistan exceeds the sustainable supply and limits our ability to provide ready forces for other contingencies ... Current

Table 7.2 US global troop deployments

	1950	1960	1970	1980	1990	2000	2003	2005
Japan	136,554	83,387	82,264	46,004	46,593	40,159	40,519	35,571
Korea	510	55,864	52,197	38,780	41,344	36,565	41,145	30,983
Germany	97,820	232,256	202,935	244,320	227,586	69,203	74,796	66,418
United States	941,231	1,565,763	1,573,500	1,266,125	1,138,627	938,753	974,571	894,430
Afghanistan	4	23	7	6	—	—	9,700	19,500

Source: Heritage Foundation. www.heritage.org/research/reports/2006/05/global-us-troop-deployment-1950-2005. Accessed June 1, 2011.

operational requirements for forces and insufficient time between deployments require a focus on counterinsurgency training and equipping to the detriment of preparedness for the full range of military missions" (2009, p. 5). Despite these stresses the report continues by noting that the US army is an "expeditionary force – organized, trained, and equipped to go anywhere in the world on short notice, against any adversary, to accomplish the assigned mission, including the ability to conduct forcible entry operations in remote, non-permissive environments" (2009, p. 11). The language of the Army report contains many euphemisms: it identifies the US army as a global force with the ability to invade the sovereign spaces of others. Despite the acknowledged stress on Army manpower, a global mission is still seen as necessary. If these statements were to be interpreted through Modelski's model, it would appear that the US military was once organized to police an established "order," but in the face of challenge the nature and location of threat has become less predictable.

Let us remind ourselves that Modelski's model is not a crystal ball. We cannot utilize its simplification of history to predict the future. However, it can be used as a framework to interpret events, the political debate after Bin Laden's death regarding the pace and size of US troop withdrawal from Afghanistan being a case in point. Is the idea that the US may suffer from imperial overstretch passé? If the United States were to follow the same cyclical pattern as previous world leaders, we would expect it to be an increasing problem.

Interpreting agency within Modelski's world leadership structure: contextualizing geopolitical codes

The purpose of introducing Modelski's model is to provide a structure within which we can situate or contextualize geopolitical agency. To do so, and in the process evaluate how effective or useful Modelski's model is, we will interpret the geopolitical codes of different states within the dynamics of the US cycle of world leadership.

The United States, the Cold War, and Modelski's model

A sketch of American history is useful to help you relate the abstract model to the "real world." The period of global war in this particular cycle ran from about 1914 to 1945, the beginning of World War I to the end of World War II. The US played a minor role in the former conflict, while it came in late and decisively in the latter. At the end of World War II, the US was able to set a global agenda around the twin themes of national self-determinism and development that established its position as world leader. Institutions such as the IMF, the UN, and NATO were established to enforce and legitimate the new world leader's agenda. However, dissent towards the US's leadership emerged, and much quicker than in previous cycles. The Soviet Union provided an immediate ideological and military challenge. The Vietnam War exposed the world leader to allegations that it supported continued European-style control of the poor ex-colonies in the world, and illustrated the limitations of its military capabilities. The Korean War and the Vietnam War are evidence that the US suffered from violent coordinated military challenge much earlier than Modelski's model would suggest. As the twentieth century drew to a close, a different form of challenge emerged at about the same time Modelski would say the US was entering the phase of deconcentration. The anti-US terrorism of Al-Qaeda had sporadic successes in Africa and the Middle East prior to the devastation of 9/11 and the heralded "War on Terror."

Broadly, the twentieth-century history of the United States fits the pattern expected from Modelski's model, though it is interesting to note that challenges to the United States' leadership came much earlier than expected, and it is a matter of both interpretation and geopolitical guesswork whether the War on Terror is a period of deconcentration preceding a new phase of global war, or the global war itself.

How can we interpret the Cold War within Modelski's model? On the one hand, the Cold War shows that the US was challenged strongly much earlier than Modelski's model would expect. The ideology of Communism, under the guise of Marxism-Leninism, offered an alternative to the liberal-capitalist model proposed by the world leader. The world leader was unable to extend its influence globally, being excluded from the Soviet Bloc and facing competition from socialist movements in Africa, Asia, and the Americas.

A key event in the era of United States world leadership was the demise of the Soviet Union and the collapse of the Iron Curtain. In a series of events through 1989 and 1991 that took commentators and policy analysts by complete surprise, the countries of Central and Eastern Europe that had been under the control of the Soviet Union since the end of World War II were allowed to renounce the Communist system. Spontaneously, in 1989, the physical barriers of the Iron Curtain, most notably the Berlin Wall, were torn down by the bare hands of jubilant people who were eager to make contact with the West. In 1991, the Soviet Union became Russia and spoke of creating a democratic political system with a market economy in place of a Communist one-party state. How should what commentators in the US interpret as the "victory over Communism" be interpreted? One argument is that the US's first cycle of world leadership was truncated and successful. The Cold War represents a victory in a Modelski-style global war that has ushered in a second consecutive cycle of world leadership for the US, under the guise of President George H. W. Bush's "new world order." However, both the lack of overt conflict with the Soviet Union and the current challenges being faced by the US undermine an interpretation that we are within the US's second cycle of leadership.

Alternative views of the Cold War may help us interpret it within Modelski's perspective. For analysts such as E. P. Thompson (1985) and György Konrád (1984), the Cold War was a mutually beneficial geopolitical drama that served the Soviet Union and the US, rather than a potential global nuclear holocaust. The Cold War provided the grounds for both major protagonists to control their allies in Western and Eastern Europe respectively. It provided the reason for the military occupation of Europe by both the Americans and the Soviets. In addition, the Cold War included a consensus that the poorer parts of the world were to be dominated by the big powers. Though both sides claimed the mantle of anti-imperialism, the Cold War provided the excuse for political and military control of the newly independent countries.

The most likely interpretation is that the Cold War signified a limited but significant challenge to the US's world leadership. In other words, the period of world leadership was muted and the period of delegitimation amplified. The argument that the Cold War was of mutual benefit to the Soviet Union and the US is supported by an interpretation that the beginning of the period of deconcentration (and not a period of stability) was marked by the collapse of the Soviet Union. All of a sudden, the certainties that the world leader had known were gone, and a violent challenge that was hard to pinpoint and counter emerged. It is to that challenge we now turn.

The War on Terror and Modelski's model

We have already spoken about terrorism and the War on Terror in the previous chapter. In this short section we will simply interpret al-Qaeda and the US's response in terms of Modelski's model. The interpretation revolves around two questions (neither of which can be answered): (1) is al-Qaeda the challenger to the United States that will drive the deconcentration and global war phases of the model, and (2) will the War on Terror weaken the US, promote imperial overstretch, and lead to its decline as world leader? The first question requires consideration that Modelski's model is about the agency of states, and if al-Qaeda (a non-state actor) were to be the challenger it would be an unprecedented development. However, the possibility should not be dismissed. Al-Qaeda certainly provides a challenge to the institutions of the world leader and its internationalist agenda. The attacks on embassies in Africa and the USS *Cole* in Yemen prior to the attacks of September 11, 2001 (and of course the attacks on US forces since 2001) are clear illustrations of the material capacity of al-Qaeda to challenge the US. That challenge is limited though, and it remains to be seen whether the organization will be able to operate effectively after the killing of Osama bin Laden.

The other side of the coin is the impact of the War on Terror on the US, and whether the invasions of Iraq and Afghanistan will drain the world leader's power in a process of imperial overstretch. Though we have discussed likely examples of imperial overstretch during the course of those invasions the trajectory is in the balance. In May 2011 (at the time of writing) Bin Laden's death had just initiated a political debate in the US as to the extent and timing of troop withdrawal from Afghanistan that could possibly reverse the US's overseas commitments. Modelski's model is aimed to give long-term, structural, and historic interpretations that would suggest Bin Laden's death will do little to change the trajectory of US decline. On the other hand, the geopolitical moment in the wake of Bin Laden's killing is an opportunity for policy changes that show the ability for agents to make choices within structural contexts.

The European Union and Modelski's model

How do we interpret the European Union (EU) within Modelski's model? First, the genesis of the EU was part of US plans to rebuild Western Europe after World War II. Though there have been political disagreements across the Atlantic since 1945, in general the US has supported the integration of Western Europe, because it helped counter the challenge of the Soviet Union and also reinforced economic and political ties with the world leader. The countries of Western Europe have, generally, followed the will of the US. One historic dispute was the British and French attempt to seize control of the Suez Canal in 1956. However, this episode met with strong US disapproval and Britain and France quickly complied with the world leader's wishes by retreating.

The EU is the product of a trend toward intensified integration of European countries, coupled with an expansion of the number of countries included. Now the EU contains the countries of Central and Eastern Europe that were once under the control of the Soviet Union. The intensification and expansion of the EU have resulted in discussions of its assumption of a global geopolitical role. In some cases this role has been visible, and in others it has been conspicuous by its absence. For example, the EU countries have been influential in international negotiations over global warming emissions. Alternatively, in the 1990s European countries stated that they would take the lead in resolving the war being waged in the former Yugoslavia, but after embarrassing failures it was ultimately the US that intervened militarily and diplomatically.

On the one hand, the EU may be viewed as a form of delegitimation, in Modelski's terms. Its growing strength and confidence have allowed some countries, notably France, to be critical of US policy. Significantly, the EU has established a military force, the EuroCorps. This may also be viewed as delegitimation: it is a statement that NATO (the military expression of US influence over Europe) is no longer taken for granted and that purely European alternatives may one day replace the world leader's institution. In 1992, the EU described EuroCorps as "a European multinational army corps that does not belong to the integrated military structure of the North Atlantic Alliance (NATO)."

On the other hand, the current EuroCorps website contains the sub-heading "A Force for the EU and the Atlantic Alliance." The EU continues to situate the role of EuroCorps within the geopolitical structures of both the EU *and* NATO. In addition, when push comes to shove, the European countries have supported the global military role of the world leader; most notably regarding the US decision to invade Iraq. The *potential* of the EuroCorps to allow the EU to project military power independent of, and even against the wishes of, the world leader is evidence of delegitimation. However, the subordination of EuroCorps within NATO, and the practical constraints on its ability to act independently of the US, is evidence of the continued power of the world leader.

In summary, the current signals from the EU are mixed: there have been verbal protests against US actions. The political decisions of the EU and institutional developments such as EuroCorps may also be interpreted as revealing discontent with the world leader's agenda. Additionally, there have been trade disputes between the EU and the US. Significantly, however, the trade disputes and construction of EuroCorps do not undermine the general agreement over free-trade policies between the EU and the US, nor the inability of the EU to define and execute military operations free from the world leader's agenda. The EU is still the world leader's key ally; though some would say an increasingly reluctant one. The documents discussed above can be found at the EuroCorps website www.eurocorps.org/, accessed May 5, 2010.

The bureaucratic complexities and economic costs of developing a European army remain (Salmon and Shepherd, 2003). However, institutional developments have continued and the Eurocorps remains involved in NATO missions. On paper, and that qualification must be stressed, the development toward a European national army continues (though at a slow pace). It has also been posited that the United States will increase its focus upon Asia, leaving political space for independent European developments, and that budget cuts for European countries will make integration and cooperation more attractive (Holworth, 2011). The November 2010 agreement between France and Great Britain for military nuclear cooperation is a significant move toward European security integration, but may also be interpreted as being driven by national financial constraints.

The European Union has been a key geopolitical ally for the US. However, there are signs that the EU is striving to gain independence from the US foreign policy agenda, while also remaining an active supporter of some elements of the world leader's geopolitical code. As the Modelski model predicts, even allies should show increasing dissent toward the leader's agenda – and one such manifestation is the tentative steps toward a European army. However, there remain many doubts as to whether such a force will actually emerge as part of the separate and collective geopolitical codes of European countries.

China and Modelski's model

The increasing global role of China has been a key feature of the changing geopolitical context in the past couple of decades. As the War on Terror has dominated the focus upon the actions of the United States, attention has also been paid to the actions of China. Three developments are of particular interest: (1) the construction of military capability, especially naval strength that increases China's ability to project power, (2) the financial power that China has gained as a result of its economic growth, and (3) its involvement on the African continent.

For some in the US military, whose job it is to envision security challenges in the not-too-distant future, China has been identified as a threat to the ability of the US to project its power across the globe, a key aspect of world leadership. In the language of US defense planners, China is developing anti-access/area denial plans that would limit the ability of the US Navy to operate freely near China's coast (Johnson and Long, 2007). Increased satellite, radar, missile, naval, and air-strike capabilities are identified as means to limit US access to part of the world's oceans. On the other hand the US and China brokered a deal to cooperate over matters of nuclear security, with an eye to the actions of North Korea as well as potential nuclear terrorism. These two different security approaches are tangible manifestations of what President Obama has called the "healthy competition" between the US and China.

In terms of Modelski's model, the discussions of area denial would point to China as a potential challenger to US world leadership. On the other hand, the move toward nuclear security cooperation suggests partnership that could develop into the coalition building that is another key part of the model. The ability for countries to develop their own agency (rather than being determined by the model's imperatives) suggests that China is not predestined to act as challenger, but that as China's power increases the US will be attempting to create both fruitful economic and diplomatic relations while, simultaneously, considering the changing security balance in East Asia.

China's rise to power is much more than a question of military might. Its staggering economic growth has meant that it has engaged with key global institutions, such as the WTO, in a dramatic shift from its traditional insular policies of the Cold War era (Moore, 2005). It has become a global financial player. Importantly it is the largest holder of US public debt, to the tune of around $900 billion. This stark fact is a source of tension between the two countries, but also means that the US has to consider the influence China has over its economy when making (ostensibly) domestic economic decisions as well as diplomatic statements about, for example, China's human rights violations.

China's increasing global profile is clear in Africa. As part of the geopolitics of the Cold War, China sent tens of thousands of workers to newly independent states. With increasing economic wealth, China was able to send investment rather than just people. The value of trade between China and Africa has grown from $6 billion in 1999 to an estimated $100 billion in 2010. Raw materials, such as oil and minerals, are sent to China to fuel its manufacturing sector and the finished goods return to Africa. This economic relationship is assisted by official Chinese aid for infrastructure projects that will benefit Chinese companies. In some cases, the financial support offered by China is used to leverage diplomatic concessions; especially promises for African states to support Chinese efforts for Taiwan to be recognized as a Chinese province rather than an independent state (Afrol.news, 2010).

China is also raising its global profile through public diplomacy. Aware of the cultural power of the US, notably the influence of Hollywood and the role of US ideas and personnel in NGOs, China is developing ways to strengthen its national image. "Whether it is Chinese philosophy, literature, aesthetics, ethics, or ancient military warfare, Chinese scholars argue that these rich and diverse traditions from one of the world's greatest and oldest civilizations should be more widely shared with and appreciated by the rest of the world" (Siow, 2010, p. 1). Some Chinese diplomats are saying that China has a need to obtain *hua yu quan*, or being given position to assert one's voice (Siow, 2010). In other words, China is developing strategies to prepare a global stage to pronounce its cultural and political qualities and achievements.

In sum, China has dramatically increased its military capability, and its economic strength and influence, over the past couple of decades. It is pursuing strategies to project cultural power, too. Modelski's model would predict that at this time some states would be positioning themselves to challenge the world leader, and of course many of the more hawkish commentators in the US are eager to portray China in this way. But as the world leader declines and faces challenges, the model suggests that it also seeks coalition partners. The complexity of dependence and suspicion that defines US–China relations suggests that there is much geopolitical agency to occur to see which, if either, of these scenarios occurs.

North Korea, the NPT, and Modelski's model

North Korea has disrupted the geopolitical agenda of the United States since the end of World War II. The Korean War (1950–3), known as the 6/25 war in Korea because of its June 25 start date, was driven by North Korea's aim to control the whole of the peninsula and was backed by the Soviet Union and China. It was the first major event of the Cold War. Since the ceasefire in 1953, North Korea has refused to compromise with the United States, providing both a justification for the United States to build up its military strength in Northeast Asia and a foil for North Korea to claim it is threatened by Western "imperialism."

The major challenge to US world leadership has been North Korea's development of a nuclear weapons program. This is not simply a military challenge, through possession of a weapon of mass destruction, but a challenge to the institutional arrangements put in place by the world leader. The Nuclear Non-Proliferation Treaty (NPT) was designed to limit the possession of nuclear weapons to the US, Soviet Union/Russia, China, France, and Great Britain (the so-called P-5 states and the permanent members of the United Nations Security Council). The NPT aimed to limit nuclear weapons proliferation by allowing states (at the outset assumed to be the P-5 states) to assist other states to develop nuclear power for domestic and peaceful purposes, such as energy and medicine, whilst stopping states developing weapons. The development of nuclear weapons programs by Israel, India, Pakistan, and North Korea is a failure of the NPT regime. On the other hand, proliferation has been limited to those states and South Africa abandoned the development of its nuclear weapons program as part of its decision to sign the NPT in 1991. Hence, a positive interpretation is that the NPT has slowed the pace and extent of nuclear proliferation. Myanmar, Iran, and Syria are suspected of trying to develop nuclear weapons programs.

To develop nuclear weapons is to challenge the ability of the US to act as world leader and enforce the NPT regime. North Korea has consistently violated the NPT by trading with other states and non-state actors to help them develop their own nuclear weapons programs. North Korea's economy is in a terrible condition and cannot provide for its citizens, with recurring famines. It has earned money by selling narcotics and counterfeiting currency (Vaicikonas, 2011). Moreover, it has traded with Iran, Syria, and Pakistan to the advancement of the nuclear weapons programs of all four states. Sanctions imposed against North Korea have been breached through trading across the border with China (Vaicikonas, 2011). North Korea's first test of a nuclear bomb in 2006 and the testing of missiles that can reach Japan are certainly a challenge to the security regime the US established in Northeast Asia. In addition, its agency in creating a trading network in violation of the NPT, and despite a system of sanctions initiated by the US, has challenged one of the world leader's key security institutions.

Legacy, change, and world leadership: feedback systems in Modelski's model

The final feature of the model we will discuss is its feedback system. Modelski identifies two related feedback systems. The first, the developmental loop, notes that though the world leaders come and go the legacy of their innovation remains. In other words, the ideas and institutions established by the world leader do not disappear entirely from the geopolitical scene as a particular country loses its status as world leader. For example, if the US was replaced as world leader it is likely that the idea of national self-determination that was an ingredient of its "innovation" will still retain some role in global geopolitics. Also, the institutions of the UN and the World Bank, as entities managing global economics and politics, are likely to remain, if perhaps in a different form. As support for this claim, the "ideas" of free-trade and freedom of movement in international waters established by world leaders hundreds of years ago remain essential political norms.

The second feedback system outlined by Modelski is the regulatory loop that examines the process of an emerging challenger and the establishment of a new world leader. The logic of Modelski's model does not allow us to make predictions. It is difficult to consider this model without asking who the next challenger will be, and who will be the next world leader. Specific answers are not provided. However, recourse to Modelski's model does raise some interesting historical patterns that help us interpret the current situation.

In Modelski's history, the next world leader has not been the challenger, but has been one of the countries in the coalition brought together by the world leader to fight the challenger. The case of the United States and Great Britain is a clear illustration of this process. Great Britain's role as world leader was challenged by Germany, resulting in the two World Wars. To challenge the might of the world leader Germany realized it needed to form a coalition; it could not do it alone. However, given the process of decline identified by Modelski, Great Britain could not fight off challenges to its power alone either. It too needed to establish a coalition of forces. Crucially, it required the industrial might of the United States to support its war effort. Germany and Great Britain, challenger and leader, exhausted their material capacity for power in the long phase of global war. Remote from the domestic destruction suffered by Great Britain, continental Europe, the Soviet Union, and Japan, the United States gained ideological influence in relation to the relative and absolute increase in its material power. Both previous leader and challengers were spent forces, but the US, the increasingly prominent member of the world leaders' coalition, was able to assume the preeminent geopolitical position. If there is a lesson to be applied from Modelski's model, it is that educated guesses about the next world leader should select from the coalition, the leader's allies, and not from its challengers.

The current geopolitical situation complicates the ability to learn from Modelski's model. Modelski's historic examples are from the period when geopolitical actors were identified as competing countries. Other geopolitical actors were ignored. What of now? If we focus solely on countries, then China, the EU, and Japan, and to a lesser extent Russia, are wheeled out every now and then as "threats" to the US. But these countries are not the cause of the US's current military mobilization. Has the geography of challenge changed? Is the network of al-Qaeda the challenger to the US's world leadership? If so, what does that mean for coalition building and the process of succession?

Pros and cons of Modelski's model

Modelski's model is helpful for putting particular events into a historical perspective. Current affairs are not singular unrelated events. Rather, they are moments in broader processes and trends. Greater understanding of the event, its significance and implications, is achieved if you evaluate it within an understanding of world politics such as is offered by Modelski. Moreover, events can also be thought of as "observations" or "data." They are the "test" of the model. In other words, do the events we see in the news counter or support the trends we expect from Modelski's cycle of world leadership? Of course, the model must be thought of broadly and as an abstract teaching tool. Nonetheless, too many deviations from the expected pattern of events should lead us to challenge the model.

The model itself is also far from perfect. But that should not force us to dismiss its value out of hand. Social scientists are well aware that the theoretical tools we work with are imperfect. One of the most important concerns toward Modelski's model, and similar ones such as Wallerstein's world-systems theory, is philosophical. First is the logical problem of historical determinism. Just because Modelski has identified cyclical patterns of world leadership in the past does not allow us to predict that the demise of the US's world leadership role is inevitable or determined. Portugal's sixteenth-century history does not determine the US's twenty-first-century future. The reason for this lies in another philosophical concern: Structural determinism. The US as world leader is a geopolitical agent; it has some degree of freedom to choose its own actions. A drift toward global war is not determined; it *partially* rests upon the actions of the US.

The key word here is "partially." Proponents of structural models tend to give emphasis to the constraints that structures place on agents, in this case the structural inability of the world leader to untangle itself from increased challenges to its authority. Researchers who are more focused upon agency place greater emphasis on, say, the foreign policy decisions made by successive US presidential administrations.

Another set of criticisms toward Modelski's model reside within his conception of what geopolitics is. First, his model follows the classic geopolitical tradition of being state-centric. The geopolitical agents (leaders, challengers, and coalition members) are all countries. Second, he focuses upon the rich and powerful countries; poorer countries of the "global south" are deemed irrelevant in his system of challenge, war, and leadership. The geography of Modelski's geopolitics is limited in two senses; it sees state territoriality as the only space of politics and it concentrates on just a part of the globe.

Power is central to any understanding of geopolitics. Hence, we should be especially critical of Modelski's measure of power. One obvious question is whether sea-power is no longer relevant in an age of cruise missiles and satellite communication. In defense of Modelski, his long-term historical perspective requires a consistent measure of power, one that is as useful for understanding the sixteenth century as it is for comprehending the twenty-first. Sea-power seems to fit the bill. The essence of the model, and the definition of power, is global reach, the ability to influence the behavior of other countries across the globe. At times this requires military muscle, and as we have seen in recent US-led conflicts that still requires a naval presence. Moreover, the US military has redefined the meaning of "global reach," utilizing weapons and surveillance systems that facilitate observation of the whole globe at all times, and the ability to remotely kill people and destroy targets across the globe (see Box 7.5). Unmanned drones carrying missiles and cameras may be a long way from sea galleons, but each identifies the world leader as the country with the dominant means of exerting its power across the globe in its respective historical period.

Box 7.5 The technology of global geopolitical reach

Research and Development efforts within the US are aimed at enhancing the technological capacity for the military's global reach. The Defense Advanced Research Projects Agency (DARPA)/Air Force Falcon program is developing hypersonic flying technology "that will enable prompt global reach missions and demonstrate affordable and responsive space lift" (DARPA, 2005). The unmanned reusable hypersonic cruise vehicle "would be capable of taking off from a conventional military runway, carrying a 12,000-pound payload, and reaching distances of 9,000 nautical miles in less than two hours. This hypersonic cruise vehicle will provide the country with a significant capability to conduct responsive missions with quick turn-around sortie rates while providing aircraft-like operability and mission-recall capability" (DARPA, 2005).

In everyday language, this military robot can fly very fast, reach across the globe, bomb a target at a moment's notice, and do it again soon afterwards, or be redirected while in flight.

A related form of global reach is Geospatial Intelligence (GEOINT), a form of military power in which the discipline of geography is heavily implicated. GEOINT is the "natural marriage" (National Geospatial-Intelligence Agency, 2004, p. 13) of satellite and rocket images with Geographic Information Systems. "By combining remote sensing, precise geopositioning, digital processing, and dissemination, GEOINT enables combatant commanders to successfully employ advanced weapons on time and on target in all-weather day–night conditions around the world. Today's warfighting capabilities represent quantum improvements in precision and targeting technologies" (2004, p. 15). And the goal? Well, "by continuing to leverage innovative technology and processes with an increasingly agile workforce, NGA (National Geospatial-Intelligence Agency) and NSG (National System for Geospatial-Intelligence) members are uniquely postured to contribute to information dominance and, ultimately, achieve the promise of a more certain world" (2004, p. 17). Contemporary global reach requires "information dominance," the goal of "knowing" the world that the world leader dominates. The purpose of GEOINT is more "efficient" military operations that will facilitate a "certain" world: not necessarily "just" or even "peaceful" but "certain," a synonym for the "order" the world leader says it can provide to justify its relative power.

Figure 7.5 Unmanned military drone.

Also interesting to note is the contradiction in the measure of power and the operation of the politics of world leadership. While the power index is based upon a material measure, sea tonnage, the process rests upon ideological power, the ability of the world leader to define and implement a political agenda that is perceived to be in the interest of all. Rather than focusing upon the number of aircraft carriers and nuclear submarines, the world leader's authority rests upon the resonance of its political and cultural institutions and practices. The resort to arms is an admission that the political agenda is not being followed.

Another pertinent question that is often raised centers upon the "driving force" underlying the model's dynamics. One attempt has been to relate the rise and fall of world powers to global changes in technology and economics (Modelski and Thompson, 1995). This raises the question of how to understand the material capacity and need to possess the sea tonnage that is an integral part of his model. Economic power requires a large merchant fleet to facilitate trade. Economic power provides the public funds needed to build a military naval capacity. In addition, Modelski forces us to look at some other processes. Most intriguing is the phasing of the preference for order. It implies a generational process of forgetting the horrors of warfare experienced by many during "global war" and an increasing truculence with an "imposed order": the geopolitics of mass psychology rather than the imperatives of capitalism.

Finally, is Modelski a geopolitician or a social scientist? A social scientist should be gathering and interpreting data with an eye to avoiding the biases of social position and nationality. Geopoliticians, on the other hand, are politicians with an eye toward advancing a particular foreign policy agenda which they believe will enhance the interests of their own country relative to others. For geopoliticians, data are collected and theories are written in order to provide a seemingly objective backdrop that makes their political agenda seem "obvious" and validated by "science." Within which camp Modelski falls is a matter of interpretation. He does have a message for the geopolitical future of the US. He is also a skilled historical social scientist who has meshed impressive data collection with an intriguing theoretical model.

Summary and segue

This chapter has introduced a particular model of world politics in order to provide a means for identifying the global geopolitical context. Though Modelski's model is far from perfect, it does allow us to situate the actions of countries within a global picture of political cooperation and conflict. In other words, the model is a way to situate or contextualize the geopolitical code of states and non-state actors within a global structure. Though geopolitical agency is our focus, Modelski's model helps us consider the structural limits, and possibilities, faced by geopolitical actors. Perhaps the most important usage is in the interpretation of the role of the US, and why it appears to be facing increased and intensified opposition.

Now that we have introduced a way of thinking of a global geopolitical structure, the next chapter will focus upon environmental geopolitics as another way to consider how broad global processes relate to geopolitical agency.

Having read this chapter you will be able to:

- Define the key components of Modelski's model
- Understand the critiques of the model
- Use it to interpret US foreign policy, as a form of geopolitical agency
- Use it to interpret the geopolitical codes of other states and non-state actors

Further reading

Bacevich, A. (2002) *American Empire: The Realities and Consequences of U.S. Diplomacy*, Cambridge, MA: Harvard University Press.

An in-depth and accessible discussion of US foreign policy decisions since the end of the Cold War. The book provides a wealth of information that may be interpreted within Modelski's model, or used to evaluate the model.

Modelski, G. (1987) *Long Cycles in World Politics*, Seattle: University of Washington Press.

The research manuscript that details the model used in this chapter and the historic data used to make the case.

Taylor, P. J. (1990) *Britain and the Cold War: 1945 as Geopolitical Transition*, London: Pinter.

Uses Wallerstein's world-systems framework to provide an accessible discussion of how Great Britain, a geopolitical actor, made foreign policy choices within the geopolitical context at the end of World War II.

Wallerstein, I. (2003) *The Decline of American Power*, New York: The New Press.

The world-systems' take on the trajectory of the United States.

8

ENVIRONMENTAL GEOPOLITICS: SECURITY AND SUSTAINABILITY

In this chapter we will:

▓ Discuss the topic of environmental geopolitics

▓ Identify different ways to understand environment–society relations

▓ Introduce the term Anthropocene to explain the current situation

▓ Discuss the securitization of the environment in the Anthropocene age

▓ Explore how the environment has become a part of the geopolitical code of states

▓ Understand how environmental threats are represented

▓ Investigate the geography of resource conflicts

▓ Examine the geopolitics of oil

In the previous chapter we introduced a global geopolitical structure created by competition between states. Another global structure that frames geopolitical agency is the environment, or more accurately our planet or the biosphere which supports life. Recently, the environment has been made a topic of geopolitics by academics and social movements concerned about environmental degradation and the sustainability of the planet. Also, the environment has become "securitized" or seen as an object toward which traditional geopolitical practices must be targeted in order to provide security. Both of these approaches, by very different geopolitical actors with different goals, regard the environment as a global structure which influences geopolitical agency and is being changed by such actions. Both sets of geopolitical actors believe that the action they propose provides "security," but what is meant by security, for whom it functions, and in what manner it is attained is hotly debated. The different approaches also have different visions of the necessary construction of spaces and places, and the way they are represented. Hence, the environment has become a global geopolitical structure framing the actions of a diversity of geopolitical agents.

The securitization of the environment involves a recasting of the geopolitical codes of states. However, the global extent of environmental change means that international organizations have come to the fore as geopolitical agents. We will see that contemporary

Box 8.1 The return of the geopolitics of the Northwest Passage

The Northwest Passage is the sea route through the Arctic Ocean along the northern coastline of North America. The search for this sea route has been a feature of geopolitics since the 1494 Treaty of Tordesillas, when Pope Alexander XI divided the world between Spanish and Portuguese spheres of influence, leaving the rest of Europe searching for an ocean pathway to Asia with its promise of trade and new colonies. The belief was that access could be found through the ice-bound oceans at the fringe of the Arctic and on the northern coastline of the North American continent. Despite many attempts, often resulting in calamity, the Northwest Passage was not successfully traveled until Roald Amundsen's exploration of 1903–6. The Arctic region came back into geopolitical prominence during the Cold War when it was discovered that nuclear submarines could travel under the polar ice-cap and position themselves for a nuclear strike.

The Northwest Passage is again a topic of geopolitical calculations, but this time the interest is driven by global climate change, or what is commonly called global warming. The polar ice-cap is melting, reducing its spatial extent and the duration of the frozen winter season. The result is that the Northwest Passage has become a viable sea route and increased potential for resource exploration is being touted. Russia, the Scandinavian countries, and the EU, as well as the United States and Canada, are situating the Northwest Passage within their geopolitical codes (Lundestad, 2009). Canada's concerns include the emergence of its northern coastline as a new entry point that requires more intensive boundary policing, as well as the potential environmental consequences of shipping, including oil tankers, traveling through hazardous channels (Byers, 2009). For others, the question of access to new fishing grounds, and the consequences for further depletion of fish stocks, requires concerted action. Russia raised the stakes, in a dramatic fashion, by using a submarine to plant a flag on the ocean floor and make some dubious geological claims to the ocean floor and, therefore, the oil and other resources believed to be located there: the decreasing ice-cap is making it easier, though still very problematic and costly, to access these resources. A 2008 US Geological Circum-Polar Resource Evaluation Survey had sparked excitement about the potential oil and gas resources that could now be reaped (Dodds, 2009).

The emergence of the Northwest Passage has also produced less belligerent geopolitics, with conventions and accords designed to manage and protect the fish and mineral resources of the area and diminish tensions between states. The Arctic Council was established in 1996 as a means of regional governance – though it is mainly a means of discussion, does not consider military or security issues, and has no legislative powers (Dodds, 2009). Despite Russia's periodic sabre-rattling, it has also been a constructive participant in the Arctic Council (Lundestad, 2009). One more point: the actions of the indigenous Inuit population have made an impact upon the perspective and actions of the interested states (Byers, 2009), illustrating

(Continued)

the significant role non-state actors are playing in the emergent environmental geopolitics of the twenty-first century.

In some ways the contemporary geopolitics of the Northwest Passage illustrates established geopolitics of boundary demarcation, resource exploitation, and inter-state cooperation and competition. However, the case of the Northwest Passage also illustrates how contemporary geopolitical codes are being constructed in reaction to changing environmental conditions and how such codes are considering environmental risks and issues. Increasingly, geopolitics and the environment are intertwined, requiring a consideration of environmental geopolitics.

Figure 8.1 US nuclear submarine in Arctic ice.

environmental geopolitics is an interplay between international cooperation and the self-interest of states. The geopolitics of, on the one hand, coping with global climate change and, on the other hand, competing to control access to vital resources illustrates that environmental geopolitics is a fascinating blend of established geopolitical practices and the necessary emergence of new ones.

Humans and the environment

To start our discussion of environmental geopolitics we must recognize two things: our generation is far from the first to wrestle with the issue of the environment and the debate has always been a political one. Four approaches to understanding human–environment interactions will be introduced briefly. The consistent theme is that humans have the ability to change the environment and in doing so different geopolitical constructions are proposed as cause and solution of environmental problems. In other words, the term "the environment" has been used as a vehicle for proposing different political agendas.

Humans are to blame I

The classic statement about humans and the environment is the treatise *An Essay on the Principle of Population* by the Reverend Thomas Malthus (1970), first published in 1798. Malthus claimed that while population tended to grow geometrically (i.e. 2, 4, 8, 16, 32 . . .) the ability to extract resources from the environment grew at a slower arithmetic rate (i.e. 1, 2, 3, 4, 5 . . .). Hence, unchecked population growth would inevitably exhaust available resources. Malthus thought that the disparity would require checks on population growth. He saw two types of checks: preventative checks decreased the birth rate, and included policies such as abstinence, marrying at a later age, homosexuality, and forms of birth control; positive checks increased the death rate and included war, disease, and famine. His social position led to firm convictions regarding who was being identified by the term "population"; it was the poor, the conclusion being that if the "poor" did not control their own population growth then "natural" checks, such as starvation, disease, and even war, would check their growth, presumably as a benefit to society.

The Malthusian approach was given contemporary credence by the famous study *Limits to Growth* (Meadows *et al.*, 1974) – basically an extrapolation of population, agricultural, and environmental resource usage trends that showed an impending planetary disaster. *Limits to Growth* is more often cited than Malthus by those who wish to impose controls on human activity in the name of the environment – probably because "science" has more authority than a politically driven sixteenth-century priest. The essence of the Malthusian approach that has implications for contemporary geopolitics is an identification of a social group, a "them," that is to blame for environmental degradation or insecurity and that this group must somehow be controlled, including spatially.

Humans are to blame II

The Malthusian approach was challenged at the time of writing. The most influential critique to emerge was that of Karl Marx, as part of his general criticism of capitalism. Marx noted that it was not the absolute size of population and the resource pool that was to blame but the manner in which resources were distributed across social groups. While Malthus drew attention to the actions of the poor, Marx redirected attention to the actions of the wealthiest and their role in consuming resources while simultaneously impoverishing the majority of the population. For Marx, Malthus was simply a representation that enabled the further exploitation of the poor working classes – to see their impoverishment (especially hunger and ill-health) as a positive thing, and one that was self-inflicted. Instead, Marx argued that the exploitative nature of the capitalist system was to blame for the situation and behavior of the poor, including high birth rates which were a response to high infant mortality rates and the need to have large families to create more "breadwinners."

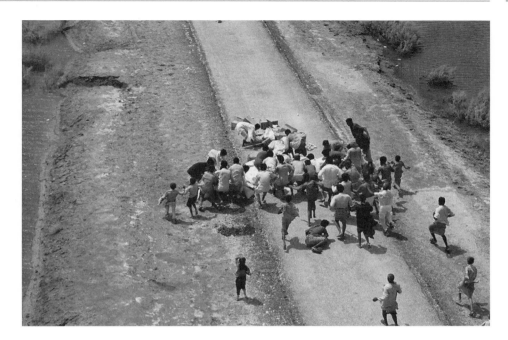

Figure 8.2 Flood relief in Pakistan.

Marx offered more than just a critique of Malthus. He also thought he had the solution in a new form of human society, Communism. In this vision, population growth was actually seen to be a good thing. As long as the population was organized along the lines of communist theory then a bigger population would be productive and provide more food, housing, and goods for the benefit of all. In the twentieth century, with the establishment of the Soviet Union and socialist China, such an approach promoted a state-led industrialization in Communist countries that paid little heed to environmental costs as production targets to provide a new socialist society were sought. The environmental legacies still exist.

In general, the message of Marx is that it is not population growth per se that should be seen as a problem but the unequal distribution of resources that allows for mass consumption by some and poor environmental practices by those just trying to survive. A starkly unequal society leads to practices of unsustainable resource exploitation and environmental practices at both ends of the social spectrum. Though centralized state Communism, as in the Soviet Union, created its own host of environmental, let alone social, problems, the Marxist approach should not be easily dismissed. The basic message that unequal and unjust resource use is at the heart of environmental degradation and geopolitics is an essential component to understanding the contemporary issues we face.

Humans may hold the solution

An alternative, and more promising, approach comes from the ground-breaking work of the anthropologist Ester Boserup who noted how the people she was studying in less

Box 8.2 The geopolitics of environmental innovation

The message that can be taken from Ester Boserup's (1965) theory can be reassuring and hopeful. It is tempting to see the massive environmental issues facing our planet as being solvable as long as the right technological fix is applied; and applied it surely will be, seeing how innovative human beings are. However, innovation must be seen as a matter of geopolitics: questions must be asked about the politics behind the application of new technology. Who gets to benefit and what are the drawbacks? More precisely, we live in a capitalist world in which ownership of innovations is held by particular companies who restrict the usage of technology to maximize profit. Innovation is about profit and costs, and not, primarily, benevolent acts to save the planet. However, in the name of making money some good can be achieved, but at some costs.

The types of innovation Boserup identified were small-scale "hands-on" behaviors in which the innovator and the user were the same group. Now innovation is big business. Hence, Boserup's framework must be considered along with Marx's: innovation is embedded within power relations between businesses that own the technology and those that must pay to use it.

This power relation is very clear within the geopolitics of food. In the 1960s a wave of innovation known as the Green Revolution was touted to end global hunger. Irrigation systems, high-yield plants, fertilizers, and pesticides were seen as increasing agricultural production. Though many people benefited, today billions of people still suffer from hunger and malnutrition. Some say that the Green Revolution benefited only a few farmers and created dependencies upon water and chemicals that were bad for the environment and led to economic reliance upon big business.

With existing large-scale hunger and projected population growth, a new wave of innovation based on biotechnology, or the Gene Revolution, is championed by some in an echo of the Green Revolution. However, there are certainly concerns about the Gene Revolution as laid out in a 2004 report *The State of World Food and Agriculture 2003–2004* by the Food and Agriculture Organization (FAO, 2004) of the UN. On the plus side, genetic engineering may produce pest-resistant and drought-resistant crops, and crops that can be grown on marginal lands but contain a high nutrient level. Though the technology has changed, the promise of better crops reducing hunger is the same message heard during the Green Revolution.

There is one very important difference between the Green and Gene Revolutions. The innovations of the Green Revolution were largely led by the public sector, but in the Gene Revolution the technologies are being driven by the private sector with an eye to making profit in the large markets of the rich countries rather than addressing the needs of the very poor and hungry. The FAO's report notes that the new research is not considering the crops that constitute the bulk of the food supply

(Continued)

of the world's poor: cowpea, millet, sorghum, and teff as well as key food crops such as wheat, rice, white maize, potato, and cassava. More generally, biotech-nological innovation is giving relatively low attention to traits with the greatest potential benefit to the poor, such as drought and salinity tolerance, disease resistance, and enhanced nutrition.

A critical analysis of the Gene Revolution raises questions about the geopolitics of new environmental technologies. Innovation is not geopolitically neutral. Political and economic power structures control how innovations are developed and applied, who gains access to them as users, and who makes strategic decisions about their availability. Especially when the private sector is the driving force, economic questions of profit and monopoly will be imperative. Technological innovation must be considered as operating within existing geopolitical structures of power and inequity.

developed countries made adaptations to their agricultural practices when faced with particular problems. In other words, human beings have the ability to innovate and problem solve. Boserup's (1965) findings have been applied more generally and we now see them in the search for new technologies that will ameliorate, or at best simply solve, a host of environmental problems. At its extreme, and this is certainly not the original intent of Boserup, such an attitude fosters a belief in the "technical fix" – or the hope that some scientists somewhere will make all the problems humans have caused go away. Behind such an assumption is the hope, perhaps even "faith" is an appropriate term, that we will not have to change our lifestyles or the way we organize society – technology will be found to allow us to keep behaving the way we have been. Representations of technical fixes usually portray life as better in the future. Hence wind power and electric cars, for example, are portrayed as ways in which suburban lifestyles based around car usage are actually good for the environment. The danger of this usage is that the very real geopolitical structures and practices related to environmental change are seen to be irrele-vant, and in fact consumption (done in the "right" way) is actually the solution.

The problem is a human one

The scientific identification of the increasing temperature of the atmosphere has refocused the way we think about the environment in general and environmental security in particular. The approaches of Malthus and Marx assumed a separation between humans (or society) and nature or the environment. But now, with a focus on global climate change, or, more commonly and inaccurately, global warming, it has become recognized that society is one of the mechanisms or components of nature: humans are part of the environment and our actions are fundamental in rapid and dramatic environmental change. Society does not impact nature, and nature (in the form of hurricanes, for example) does not act upon society, but human society is one causal agent driving environmental change.

Box 8.3 What is global climate change?

The term global climate change is the scientific term used for what has become known as "global warming." The premise rests upon the amount of carbon dioxide in emissions from the burning of carbon or fossil fuels (such as coal and oil). The carbon is released into the atmosphere, where it prevents solar energy naturally bouncing off the earth from entering space. The energy is trapped and causes the temperature of the atmosphere to rise; hence carbon dioxide and other related emissions are known as "greenhouse gasses." Scientists have investigated a "carbon balance" in our biosphere – or the way that carbon is captured or sequestered naturally – such as in the oceans and, importantly, the forests. Hence, deforestation is a key component of global climate change as it reduces the amount of trees able to "recycle" carbon dioxide into oxygen through evapotranspiration and also destroys natural "carbon stores" and releases that carbon into the atmosphere as a gas.

Scientific analysis has indicated that the temperature of the planet changed dramatically in previous geological periods, but that such change was gradual and not caused by humans; though it led to severe environmental changes, such as species extinctions. However, scientists are in overwhelming agreement, with high levels of confidence, that the temperature of the planet has risen quickly since the Industrial Revolution as a result of human usage of carbon-based fuels and the consequent release of carbon dioxide into the atmosphere.

Though this has always been the case, it is poignant to reflect how and when the form of human action as part of the environment changed fundamentally to become a mechanism that is causing drastic change. The commonly held answer is the Industrial Revolution of the late 1700s, or humans' ability to use carbon-based fuels (especially coal, initially) in enormous quantities to create energy. The technology of coal, such as the steam engine and, subsequently, coal-based power stations, was followed by the ability to use oil. Oil not only fuelled industry but, with the invention of the internal combustion engine, led to the mass usage of the automobile and related processes of urbanization. This human usage of carbon fuels, and the effect of changing the nature of our atmosphere with further implications of sea-level rise and changing rainfall patterns, is causing massive environmental change. Though environmental change at this scale has happened before (i.e. Ice Ages), the fact that it is human-made means that it is happening at an unprecedented pace, with spiral effects and thresholds of change that we are unable to predict at the moment. The central role of humans in this process of environmental change, beginning with the Industrial Revolution, has led to labeling our modern times as the Anthropocene age (Crutzen, 2002; Dalby, 2011).

Recognition of the Anthropocene age requires us to rethink nature–society relations and, therefore, traditional mainstream approaches to security or geopolitics. Realizing that we humans are part of the environment nullifies the traditional resort to binaries that has dominated geopolitical thought. The basis of classical geopolitics was labeling parts of the world as "outside" or "foreign" and inherently dangerous, requiring in turn national security policies to provide protection for a particular state-defined population. Yet in the Anthropocene age there is no meaningful role of inside/outside or external agents causing threats to a national group, and only that national group. Instead, "we" are part of the security issue; our own actions are creating what has become deemed a security threat. Dalby (2011) is more specific about the identity of the "we": it is the most affluent and their consumption patterns that are the drivers of a security threat that creates risks and threats for all, but especially the poorest.

Despite the need to rethink security in the Anthropocene age it is probably unsurprising that states and their military planners have been tempted to try and tackle the new and unprecedented problem within their existing frameworks. Some commentators have been glad to represent the situation in a way that would be familiar to classical geopoliticians. In other words, the environment has been "securitized" and it is to this development we now turn.

Securitizing the environment

The move toward seeing the environment as a matter of security – and national security, more accurately – was not just caused by recognition of global climate change and other planetary processes. The awareness of human-induced environmental change coincided with the end of the Cold War and the related search by states and their militaries to find new tasks or, to put it bluntly, reasons to exist. Though the terrorist attacks of September 11, 2001 were soon to catalyze the War on Terror and a focus on militarized foreign policy, the 1990s were a time when the militaries of Western countries were searching for things to do – summed up in the wonderful and telling term MOOTWA (Military Operations Other than War).

The blend of scientific and political recognition of the implications of the Anthropocene age, a desire for the military establishment to be involved in this new arena of security, and consideration of how conflicts are, and will be, partially effected by environmental change all combined to make us see the environment through the lens of security. Though there is certainly awareness that the processes of environmental change are different from those of conflict, the response has still been framed, largely, by different national security establishments and their ingrained ways of evaluating threats and responses to those threats.

The current situation is a mismatch between a very new form of threat and established ways of approaching security. Securitization is the process by which certain entities or processes become defined as threatening or dangerous and hence require a security response, usually in a military form (Bernazzoli and Flint, 2009). Classic geopolitics has usually equated threats with other states requiring a (military) response by the threatened state. This threat–response relationship has created what has become known as the "security dilemma"; or the actions of one state deemed as being defensive from its perspective are seen as being threatening by another, requiring it to respond in a manner seen as threatening by the first state and hence creating insecurity rather than security.

This is an established way of critiquing arms races between states. However, the situation is very different when considering the environment as a security matter. As Dalby notes:

> In the case of climate change "our" actions in the developed affluent world, where security is studied, and books on security written and read in universities, are directly threatening to people in poor states who are more vulnerable, and directly threatening to future generations whose options will be drastically curtailed if nothing is done to alter existing trends in greenhouse gas production. But unlike traditional security studies, those whom we threaten – the poor in the global South, or our as yet unborn grandchildren and great-grandchildren – are not planning to defend themselves; nor do they have the ability to threaten affluent Northern states in any plausible manner. They don't have armies that can invade; they don't have navies or air forces to transport those non-existent armies either. In the long run, however, with sea-level rise, disasters increasing, and major disruptions to agriculture and the global economy, the affluent societies that have set these trends in motion will be directly affected too.
>
> (Dalby, 2011, pp. 92–3)

These are sobering words, and ones that force us to consider our own daily behavior with its environmental impacts as a form of geopolitical agency. The Anthropocene age also requires us to reflect upon whether securitizing the environment, or seeing it as an arena for national security agency, is likely to be beneficial. Daniel Deudney (1990 and 1999) is far from being optimistic. In a series of articles he has argued that seeing the environment as a matter of national security is likely to be counterproductive, because the issue is a global one requiring international cooperation. Thinking about the environment through the lens of nationalism frames the matter in terms of the actions of specific "others" (e.g. China's economic growth) rather than seeing the Anthropocene age as caused by and having implications for humanity as a whole.

Environmentalism as geopolitical code

Though scholars such as Dalby and Deudney argue that national security is an inappropriate response to global environmental issues, states have included the environment as part of their national security calculations, or geopolitical codes. In other words, a national response to what are perceived as environmental threats has become a new form of geopolitical agency. Such calculations follow the same formula as geopolitical codes focused upon states: identifying threats, devising responses to them, and justifying both.

For the United States, the 2006 National Security Strategy began to talk about environmental issues that had "no borders." Such a cliché was embedded within more traditional security concerns of terrorism, weapons of mass destruction, and regional conflicts between states. President George W. Bush's administration was resistant to recognizing global climate change; hence the environment was identified as "natural disasters" such as hurricanes and tsunamis. Also, the US portrayed itself as a benevolent geopolitical agent that would send aid, in the form of an initial and emergency military

---

Activity

Dalby's (2011, p. 50) discussion of the tension between national security and global environmental processes highlights a key quote from Deudney (1999, p. 214):

> The movement to preserve the habitability of the planet for future generations must directly challenge the power of state centric nationalism and the chronic militarization of public discourse. Environmental degradation is not a threat to national security. Rather, environmentalism is a threat to the conceptual hegemony of state centered national security discourses and institutions. For environmentalists to dress their programs in the blood-soaked garments of the war system betrays their core values and creates confusion about the real tasks at hand.

To what extent is Deudney's criticism of environmentalists fair? Greenpeace is one of the most prominent environmental movements. A Google search quickly identifies two different websites: Greenpeace USA and Greenpeace International. In my home in the US if I try to go to www.greenpeace.org my browser automatically takes me to the group's US site – or in other words by default it nationalizes my situation. Does this matter? Are there any significant differences between the US (or any other national) site and the international one? Both sites use the word "we" to describe Greenpeace and its actions. Does the meaning of the word "we" alter depending upon whether you are reading the international or a national website? Is environmental action more or less effective if it is organized nationally? Is it inevitable or unavoidable for an environmentalist movement with a global perspective, such as Greenpeace, to organize as an aggregation of national groups?

---

response, that reflected (and justified) its global geopolitical presence while identifying other states as weak.

Since 2006 the foreign policy establishment in Washington, DC has begun to take a more concerted approach to the environment. In 2009 the CIA launched The Center for Climate Change and National Security, a joint operation of the Directorates of Intelligence on the one hand, and Science and Technology on the other; an example of securitization through combining scientific analysis of environmental processes with geopolitical evaluations. The press release claimed: "Its charter is not the science of climate change, but the national security impact of phenomena such as desertification, rising sea levels, population shifts, and heightened competition for natural resources. The Center will provide support to American policymakers as they negotiate, implement, and verify international agreements on environmental issues. That is something the CIA has done for years" (CIA, 2009). In other words, though the environment is recognized as a global issue, the role of the CIA is to translate it into national concerns and actions that are perceived to be for the benefit of the US. The traditional national-centric perspective of a geopolitical code is unchanged.

The CIA press release goes on to say that the Agency will be working with universities and think tanks to address issues of environmental security. A project of constructing geopolitical knowledge, in the tradition of classic geopolitics, is at hand. There are numerous examples of this. One is a recent volume entitled *Climate Change and National*

Figure 8.3 Military response to environmental risk.

Security: A Country-Level Analysis (Moran, 2011). Tellingly, the editor, Daniel Moran, is a Professor at the Naval Postgraduate School and the publisher's description of the book claims it focuses upon how "environmental stress may be translated into political, social, economic, and military challenges in the future." In other words, the environment is just one more calculation to be plugged into a state versus state geopolitical analysis that is the building block of a geopolitical code.

Of course, the United States is not the only country developing its geopolitical code with an eye toward the environment; though its status as world leader would, in Modelski's (1987) view, lead to expectations that it will be attempting to set the environmental security agenda. Other countries have included environmental issues in their agendas, meaning that the environment has become a ubiquitous issue in the construction of geopolitical codes. For example, tensions between India and Bangladesh that originate from the dissolution of British India in 1947 are heightened by environmental issues. Ali's (2005) discussion focuses upon the role of deforestation, increased sediment loads in rivers, and the extraction of water from the Ganges River by India, to the detriment of Bangladesh.

Ali (2005) highlights the ongoing border dispute between India and Bangladesh, and its intersection with the environment around the issue of boundary demarcation, akin to the issues we saw in the Hypothetica case in Chapter 5. First, changes in river patterns led to newly available farmland in the Belonia area of the India–Bangladesh border. Under the protection of the Indian Border Security Force, Indian farmers grew crops on 50 acres of newly exposed land along the Muhuri River. The changing environment provided opportunities for farmers, but this became a matter of national competition and military action. Second, environmental change led to the appearance of a new island by the mouth of the Hariya Bhanga River in 1971. The island was claimed by both India and Bangladesh, compounding a long-running dispute over the course of their maritime boundary (Ali, 2005). Environmental change was a component of tensions over the geographic extent of India and Bangladesh.

The problems Ali (2005) identifies in the case of India–Bangladesh are applicable to a host of countries and regional contexts. They reflect a sense that environmental issues have security implications, and that traditional security concerns (such as the course of a boundary) are influenced by environmental change. However, such a geopolitical perspective does not reflect the changing understanding of human–environment relations in the Anthropocene era (Crutzen, 2002), or the subsequent need to rethink security (Dalby, 2011). Instead, the analysis of the environment in this way leads to seeing it as just another element in a national security calculation, as if it were equivalent to a neighboring country's acquisition of a new aircraft carrier: the environment is just another issue the military needs to think about. However, though the inclusion of the environment in nationally based geopolitical codes is pervasive, it is not the only way states are acting geopolitically because of environmental concerns. States have included cooperation with other states as part of their geopolitical codes. It is to the explosion of inter-governmental bodies and agreements that we now turn.

The necessity of inter-state geopolitics?

Environmental issues have a long history of requiring, or facilitating, inter-state cooperation. Though some traditions see the environment in terms of resources that are the focus for competition or conflict between states, inter-state agreement is quite common. One of the best examples is the continent of Antarctica, recognized as a "natural reserve, devoted to peace and science" in a 1959 treaty. It is the one swathe of territory on the globe that has been deemed off limits to sole claims of national sovereignty. The establishment of international oceans rests on the same principle and has become a taken-for-granted part of our political world. Shared freshwater resources, or river basin management, are another example of inter-state cooperation (Harris, 2005). However, regular warnings about the likelihood of future water conflicts show that inter-state cooperation over environmental issues is a geopolitical *process*, constantly requiring negotiation.

The Anthropocene age has ushered in a series of conferences, workshops, protocols, and agreements between states. In this section we will emphasize the positives and potential of such inter-state cooperation, with an emphasis upon global climate change, before introducing some critical reflections in the next section. The UN has acted as a framework to organize an inter-state response to global climate change. The Intergovernmental Panel on Climate Change (IPCC) was established by a resolution of the UN General Assembly in 1988 after the establishment of the Panel by the UN Environmental Panel and the World Meteorological Organization (WMO). The IPCC came to global prominence when it was awarded the Nobel Peace Prize in 2007. The IPCC website (www.ipcc.ch/organization/organization.shtml) is pregnant with the language of inter-state cooperation. For example, "The IPCC is an intergovernmental body. It is open to all member countries of the United Nations (UN) and WMO. Currently 194 countries are members of the IPCC. Governments participate in the review process and the plenary Sessions, where main decisions about the IPCC work programme are taken and reports are accepted, adopted and approved." The website goes on to emphasize the support of governments for the scientific endeavors of the IPCC, but then makes one thing clear: "Because of its scientific and intergovernmental nature, the IPCC embodies a unique opportunity to provide rigorous and balanced scientific information to decision makers. By endorsing the IPCC reports, governments acknowledge the authority of their scientific content. The work of the organization is therefore policy-relevant and yet policy-neutral, never policy-prescriptive."

The establishment and remit of the IPCC, and its award of a Nobel Peace Prize, are indicative of a number of key points. First, global climate change is recognized by individual governments and the UN as a problem requiring attention and action. Second, the issue of global climate change has produced inter-governmental action to staff and support the IPCC. Third, the actions of the IPCC, and therefore the issue of global climate change, have been given visibility and legitimacy as a "global good," something that is a moral imperative of the international community. Fourth, despite these positives, the impact of the IPCC has been, and will likely remain, limited. The key position of states as geopolitical actors means that action to combat global climate change will not take place unless states can reconcile this concern with other elements of their geopolitical code. There is also one other factor: engaging global climate change requires concerted global geopolitical action – a task that would appear to fall under the responsibility of the world leader. Is the United States willing and able to make global climate change an important part of its geopolitical code?

International responses and national geopolitical codes

Though an institution like the IPCC can act above states, produce knowledge, and make recommendations, it is states that must act to implement regulations and practices to limit fossil fuel emissions. A series of international conferences and workshops has produced protocols and targets for fossil fuel emissions by states. This has led to a battle in which the rich industrialized countries (the biggest emitters) are trying to minimize the reductions they must make, and the developing countries argue that they are most vulnerable to climate change and should not be punished for the behavior of the major emitting states. In addition, rapidly growing countries such as China and India are wary that their trajectory will be delayed by having environmental conditions imposed upon them. In sum, competition between states and their position within the global economy produces a geopolitics of negotiation in which an international response to global climate change becomes a matter of state-based geopolitics.

The Kyoto Protocol is perhaps the best example of the tension between a global vision and the goals and imperatives of individual states. The Kyoto Protocol refers to an agreement stemming from a meeting in Kyoto in December 1997 under the umbrella of the UN Framework Convention on Climate Change (UNFCCC). The UNFCCC website has more details: http://unfccc.int/kyoto_protocol/items/2830.php. The Protocol set binding targets for thirty-seven industrialized countries plus the EU to reduce greenhouse gas emissions. This amounts to a 5 percent reduction from 1990 levels through the five-year period 2008–12. However, we can see the operation of what we identified in Chapter 1 as relational power in the actual operation of the Kyoto Protocol.

The agreement recognized the unequal power relations between the richest countries and the rest of the world in the principle of "common but differentiated responsibilities." The principle accepts that global climate change is the result of the industrialization over the past two centuries of the world's richest countries. Hence, they should carry the burden of reducing the process of climate change. The Kyoto Protocol identifies this burden or responsibility in two ways. The first is by demanding that the industrialized countries reduce emissions. The second is through mechanisms of what has become known as carbon trading (so named because carbon is the most important element in greenhouse gasses). The protocol set emission amounts for each country. If a country emits less than the set amount it may sell that amount to a country that has exceeded its target of emissions. In other words, a dominant position in the network of global economic relations gives a country the ability, or power, to pay for emitting over its agreed limits.

Though the convention to establish the Kyoto Protocol occurred in 1997 it did not enter into force until 2005. However, the Protocol was shaken by the controversial decision of President George W. Bush in 2001 to withdraw from the agreement. He argued that the Protocol was flawed as it did not require developing countries to reduce their emissions, including the relatively powerful and rapidly growing countries, India and China. President Bush's overall conclusion was that the Protocol would harm the US economy. His argument is very state-centric, that a country's geopolitical code must be about narrow self-interest. In contrast, the principle of "common but differentiated responsibilities" takes into consideration the history of all states, how the challenges and insecurities of global climate change that effect everyone have been constructed over time by the actions of a minority of states, and how these very actions have produced their relative wealth and power. President Bush acted in a traditional geopolitical practice of state competition and self-interest. But does global climate change perhaps require geopolitical codes that create common responsibilities and actions?

The attitude of the US to global climate change remained state-centric with the actions of President Barack Obama's administration. The 2009 Copenhagen Climate Conference was part of the UNFCCC program and was intended to come up with agreements to identify actions to mitigate climate change beyond 2012. The media claimed the conference ended in "disarray," with an Accord (rather than a concrete agreement) that was not agreed upon unanimously. Subsequently, and through the channels of Wikileaks, it became evident that China and the US had acted in concert to ensure that no agreement

Figure 8.4 Climate scientists conducting research.

was reached. In other words a mutually convenient relationship of cooperation between two states was able to counteract attempts to act in a communal manner. It appears that geopolitics of state interest, in which powerful states can attain their goals, continues, even in the face of a challenge that will affect all of humanity. If that is the case, then non-state geopolitical agents are likely to play an increasingly important role.

Climate change and non-state geopolitical agents

The role of power politics in disrupting inter-state agreements on global climate change suggests that non-state geopolitical agents may be more apt at identifying the needs and reaching the goals of those who will be most affected by global climate change. Though the change is global the impact will fall disproportionately on the poorest within the poorest countries. The Earth Summit, held in Rio de Janeiro in 1992, was an unprecedented UN conference that brought together 2,400 representatives of NGOs including delegates from groups representing indigenous and marginalized peoples in developing countries. The Summit produced agreements and frameworks, including the UNFCCC, the Statement of Forest Principles, and the UN Convention on Biological Diversity. A follow-up conference (Rio+20) is scheduled for the same venue in 2012.

One significant geopolitical implication of the Earth Summits is the recognition of non-state agents as essential actors. The Summits enable diverse groups from across the globe to forge connections and make joint statements and actions. These connections show a promise of common responsibility that is different from the interpretation and actions of states after the Kyoto Protocol. In addition, social movements such as Greenpeace are able to articulate the insecurities of environmental change in a very different way than states do. In sum, the geopolitical representation that states provide security is under-mined when the threat is one common to humanity but generated, largely, by the most powerful. The delegitimation of states within the context of environmental change provides a context for the increasing role of the types of social movements we discussed in Chapter 6. However, the sense of connection and common responsibility that social movements have been successful in generating is still counteracted by dominant geopolitical representations of threat and otherness. An environmental geopolitics that considers the security of humanity must be able to identify and critique these dominant representations, the focus of the next section.

The return of Malthus

The Reverend Thomas Malthus has had a lasting impact, and is readily mobilized in geopolitical representations that are eager to do two things: (1) blame people other than ourselves for environmental degradation and (2) identify environmental issues as a security threat that must be incorporated into geopolitical codes, usually with military involvement. Contemporary uses of Malthus, known as neo-Malthusian in the literature, are based upon geopolitical representations that concentrate on the global and unbounded nature of new environmental dangers (Dalby, 2006). In doing so a dualism between society and environment is maintained, some connections across the globe are emphasized, but at the same time other linkages are denied.

The most influential neo-Malthusian text was Robert Kaplan's (1994) essay "The Coming Anarchy" which showed a global South rife with environmental degradation, starvation and hunger, and disease (AIDS was highlighted) that were a function of unchecked population growth. It was vintage Malthus but with a global, or classic geopolitical, range. Kaplan specifically identified the impoverished masses being pushed to challenge the more prosperous North, building upon images of a wave of environmental refugees moving through Central America and crossing the Mediterranean Sea to seek survival and protection.

The geographic understanding of the world to provoke such a scenario – and it was a view eagerly consumed in the wealthier countries – required a carefully crafted geo-political representation. First, the environment is seen as being separated from human

Box 8.4 Water wars?

One focus of neo-Malthusian geopolitics has been water. In 2005 ex-UN General Secretary Boutros Boutros Ghali made a statement in an interview with the BBC that competition for water resources in Africa and the Middle East could lead to war (BBC, 2005). This statement was one of a series that led to the idea of water wars – or conflict about access to potable water. A consideration of the geopolitics of water requires us to think of a new geopolitical scale: the watershed. Rivers do not conform to international boundaries. Rivers connect countries but can also be a source of tension. An upstream state can halt or curtail the flow of water through dam projects and over-usage. It can also pollute the water to the further detriment of a downstream state. Usually upstream states have geopolitical power over downstream states, such as in the Mekong River basin (with Chinese dams allegedly causing water shortages to the downstream states of Thailand, Laos, Burma, Cambodia, and Vietnam). An interesting anomaly is the Nile River basin in which Egypt is a powerful downstream state that has threatened to bomb any dams that were to be built by upstream Sudan.

John Agnew (2011) has argued that the politics of water, especially the way that great rivers connect the various (and sometimes competing) interest of different states, requires an application of politics that is not conflictual or based on zero-sum games but requires consideration of the opinions and needs of others. A positive view of the geopolitics of water is that rather than provoking wars it will require a collaborative geopolitics. Furthermore, the work of Leila Harris (2005) shows that this geopolitics will require connections between local, regional, state, and international scales.

However, whether it is to be future water wars or collaborative watershed geopolitics is a matter of geopolitical agency. International agencies may provide an institutional context or structure to nurture cooperative agreements over water resources rather than wars. In other words, water can be seen as a conduit to peace rather than conflict if geopolitical actors can be guided toward such possibilities. The International Hydrological Programme within UNESCO has established an initiative called From Potential Conflict to Co-operation Potential (PCCP) that "facilitates multi-level and interdisciplinary dialogues in order to foster peace, co-operation and development related to the management of shared water resources." Rather than the dire prognosis of a neo-Malthusian viewpoint, institutional Boserupian innovation may foster peaceful collaboration around shared scarce resources.

See the PCCP website at www.unesco.org/water/wwap/pccp/ for more details. Accessed May 26, 2011.

activity; it becomes something external or a new type of "other" that has a dangerous impact upon innocent humans (Dalby, 1996). Disease, environmental degradation, and global climate change are not seen as social processes or the result of human activity. This is clearly an incomplete and biased perspective. Deforestation and desertification are the result of human actions, such as clear-cutting of rainforests to harvest timber. It is like claiming that nuclear proliferation is independent of the decisions of state leaders.

Second, certain geographic connections, some of them implausible, are represented as being likely (Dalby, 1996). This is a common component of geopolitical representations, constructing scenarios that bolster a weak argument and have implications that demand a security response. For Kaplan (1994) it was the movement of people from global South to global North. This exaggerated threat denies the necessary ties people have to place that sustain them even in times of hardship. Though massive disruptions, including war and natural disasters, do provoke large-scale movements of populations, these events are rare. The components of place we identified in Chapter 1, including a source of livelihood, supporting institutions, and a sense of belonging, create bonds to places that make the cost of leaving high. In addition, the costs of moving in the conditions described by Kaplan would be enormous; being identified as a security threat would lead to the exclusion of environmental refugees from everyday society. They would be isolated in camps and likely be sent back across international boundaries.

Third, and in contrast to exaggerating weak or unlikely ties, Kaplan's analysis ignores actual and important connections between the global South and North (Dalby, 1996). The environment and populations of the global South are seen as an isolated "other." In reality, the environments being degraded are connected to the global North through linkages of investment and trade. Simply, the timber harvested in the forests of Brazil or Burma is consumed as furniture or other products in the stores of the US, Europe, Japan, China, etc. The pollution of the Niger delta and the poverty and violence experienced there are a direct result of oil exploitation by multi-national companies and the demand for oil products in high-consumption societies. When discussing Massey's (1994) understanding of place in Chapter 1 we noted the importance of identifying the connectivity between places; and investment and trade would be one such connection. Furthermore, while powerful actors may create or encourage some linkages (such as those involved in oil production) others may be discouraged. The countries of the global South have consistently argued that the best way to combat the poverty that Kaplan laments is to end trade tariffs that protect US, EU, and Japanese farmers from competition.

As we have emphasized throughout the book, representation is an essential component of geopolitical codes. This does not change when it comes to environmental geopolitics, and the specter of masses of the poor as somehow dangerous to the lives of the comparatively wealthy and powerful has been updated from the time of Thomas Malthus. The ubiquity of neo-Malthusianism is not surprising. It tells a comforting story: the poor are to blame for their situation, they are destroying the environment, and security actions against them are justified. This representation helps the comfortable and wealthy (including me and, I suspect, you) sleep at night. The power of a geographic perspective, such as Massey's (1994), that highlights real, rather than represented, connections between places is the establishment of a geopolitical awareness of how one group's poverty could be a function of our wealth; and it is "our" consumption that drives "their" environmental degradation. That is a much more unsettling story – though one more likely to incite collective geopolitical actions that may have positive impacts on global environmental change.

As we have emphasized throughout this book, geopolitics is a matter of practice and representation. The representations discussed in this section are motivated by a sense of limited resources and competition over their control. In some cases such competition actually results in war, and it is to the practice of resource wars we now turn.

Territory, conflict, and the environment

The state-centric geopolitical approach to the environment rests upon the classical geopolitical tradition that aimed to improve the relative position of one particular state. The national security agenda of geopolitics has always had an environmental angle, usually the identification of key resources that need to be controlled by a particular state. The term resource becomes represented as a strategic resource, which implies it is something with more than simply financial or use value, but is necessary for the safety of the nation. A resource then becomes nationalized as something that a particular nation should control, either by ensuring access through the markets or by controlling the territory within which the resource is located.

The identification of something as a strategic resource will change over time. At the time of British naval power toward the end of the nineteenth century the strategic need was for global access to coal in order to power the battleships of the Royal Navy. A network of coaling stations was established across the globe to ensure that the fleet could be refueled and therefore enable the global reach of Britain (Harkavy, 2007). Through the twentieth century, coal was replaced by oil as the key strategic resource, both to fuel domestic economic production and to the sophisticated war machines of the powerful states. Another trend is the increasing commodification of resources, or in other words making things that exist naturally into tradable things with financial value. Land and water are prime examples. Land has become real estate, something bounded and identified as belonging to a particular individual, business, or state. Water is a "natural resource" that is increasingly privatized, and hence access to it is controlled.

The geopolitical practice of resource control can be traced throughout history. In our engagement with modern geopolitics we can see the imperative in the early expeditionary activities of modern states, and the way in which they sponsored expeditions to map out the "unknown" world and claim part of it as theirs. This imperative for territorial control underlay what Agnew (2003) called "civilizational" geopolitics – or the process by which powerful European countries extended control across the globe. Explorers reported back to their sponsors "new found" lands and the riches they contained, often resulting in the territory being claimed as a colonial possession. Through such a process, and the related process of state building, the territorial component of the environment became compartmentalized into territorial units claimed to be under the sovereignty of a particular political entity. The environment was not something under the stewardship of indigenous communities but was controlled by state and external colonial powers. It became part of the calculations of global power politics, and that is how it remains.

Exploration to control resources required a balance between private entrepreneurs seeking wealth through trade and states seeking power through territorial control. Famous explorers such as Sir Francis Drake and Christopher Columbus were sponsored by national royalty with an eye to establishing trade that would establish their personal wealth as well as the tax coffers of the royal court. Though the form of both politics and enterprise has changed considerably, the same basic relationship exists today. For example, criticism of the invasion of Iraq in 2003 included claims that the purpose was to establish control of oil deposits that were to be tapped and controlled by multi-national companies with particular links to national governments.

Attention has turned to resources as a cause of war, or so-called resource conflicts. If resources are something with monetary and strategic value then they will be the targets of attempts to control them through different forms of politics. More specifically, inquiry has focused upon the relationship between a country's dependence on resources and its tendency to experience conflict, usually civil wars. Using quantitative analysis, political scientists have shown a relationship between important resources and civil war (for a review see O'Loughlin and Raleigh, 2008). The geographic perspective raises caution about making such general claims. Instead, geographer Philippe Le Billon (2005) has stressed the importance of the role of geographic context in bringing together different causes and outcomes in different geographic settings. In other words, relationships around resources that produce a civil war in one country might not produce conflict in another, and wars may be related to resources differently in different countries.

The general reasons for connecting wars with resources have led to the label the "resource curse" (Le Billon, 2005). Tracing the routes back to the time of early imperial conquest, Le Billon notes the connection between war, trade, and power centered upon overseas conquest for resources. By the end of the nineteenth century, industrialization had increased dependence on resources located overseas to such a degree that formal empires were thought necessary to ensure access. In the first half of the twentieth century the rapid increase in technologies requiring oil created new dependencies, and these were sometimes entwined with the politics of the Cold War as the US and the Soviet Union competed for access to states with oil reserves. Iran and Iraq are both good examples of states that were courted by Cold War powers to be "client" states that would allow access, and favorable deals, for "national" oil companies.

With growing concern over the environmental health of the planet, a new linkage between resources and conflict was made and epitomized by the term "green wars." The idea was that processes of environmental degradation (such as soil erosion, deforestation, drought, etc.) would create high levels of social stress that would lead to the outbreak of war. The logic behind the idea of "green wars" comes from Malthusian thinking – especially the relationship between scarcity of resources and population growth. Marx's original critique of Malthus can be adapted to challenge a simple connection between scarcity and war. Markets for resources are global rather than local, and hence scarcity is a function of where resources are consumed to satisfy a global consumer base. This is not a new phenomenon. Mike Davis (2001) has noted the existence of "late Victorian holocausts" at the end of the nineteenth century; or famines in places that were simultaneously exporting food. The construction of scarcity is not a local matter, or one in which we can simply bound a "local" population and "their" resources that "they," apparently, are degrading. Scarcity is a matter of being able to buy resources on a global commodity market. A related emerging issue is the increasing attention being paid to biofuels as an alternative source of energy to carbon, or fossil, fuels. The potential is for an increase in the price of corn as it becomes a fuel source rather than a food staple, decreasing the ability of some to afford to eat so that others can continue to drive.

An alternative approach to resources and conflict emerged after the Cold War as wars in and between what had been client states of the two superpowers broke out, particular focus being given to wars in Africa. States that had received support from either the United States or the Soviet Union for strategic reasons during the Cold War were left to their own devices. In response they turned to controlling resources within their borders in what became labeled "greed wars." The key change here is that resources switched from being the reason *for* war to becoming the *means* and end of war. The purpose of controlling resources, especially diamonds and timber, is to pay for waging war. The outcome of war is the ability to maintain control of the territory in which the resource is located. These "new wars" are about profit, and certainly not ideology or traditional causes of national liberation.

Le Billon (2005) recognizes the general trends identified in the terms "green wars" and "greed wars," but when it comes to understanding specific wars he cautions against simple and universal explanations. Countries that rely on the export of primary resources are also likely to experience undemocratic politics and poor economies, because they are dependent on the price of a single commodity that is determined by the global market, and the state's rulers are often able to rely on the money earned abroad and can ignore the well-being of their people. However, whether such circumstances lead to war, and if so in what form, is a function of the specific circumstances of the country.

Though it is hard to simplify the complexity of unique cases into an explanatory framework, Le Billon concludes his discussion of resource conflicts by suggesting how different types of resources and their geographic location are likely to produce particular types of conflicts. First, he categorizes resources as being either concentrated in a particular location (what he calls "point") or produced over a large area (what he calls "diffuse"). Examples of point resources include oil reserves and examples of diffuse resources include timber and cropland. The reason why this distinction is important lies in the ability to "harvest" and then move the resource to market, or what Le Billon calls the "lootability" of the resource. The nature and location of the resource relate to how it can be targeted by rebel or non-state forces. For example, on-shore oil is hard for rebel groups to actually drill or exploit, but it may be stolen or facilities may be targeted for extortion. On the other hand, alluvial diamonds are liable to be exploited or stolen by rebels.

Taking this analysis a step further, Le Billon relates the type of resource (point or diffuse) to its location within a state, whether it is close (proximate) to the centers of power or nearer porous borders and in remote, poorly governed areas (distant). The result is a 2 by 2 categorization based on resource type and location that suggests the form of violence that rebel groups are most likely to use to control the resource (Table 8.1). Point resources, such as oil and gas that are close to centers of power, are most likely to provoke a *coup d'état* to take control of the state and thus the exploitation of the resource. However, if the same point resource is distant it is more likely to lead to secessionist politics, or attempts to create a new state that contains the resource. A diffuse

Table 8.1 The geography of resource conflicts.

Resource characteristic	Point	Diffuse
Proximate	State control or coup Iraq/Kuwait (oil)	Peasant or mass rebellion Mexico-Chiapas (cropland)
Distant	Secession Chechnya (oil)	Warlordism Burma (timber)

Source: Le Billon, 2005

Activity

What do we know about the connection between environmental stress, climate change, and armed conflict? There is a general idea that climate change will heighten and intensify existing environmental difficulties and lead to wars. One rigorous test of these ideas has suggested that we should be more cautious about making definite and complete claims (Raleigh and Urdal, 2007). Instead, population growth and population density seem to be related to conflict when some types of environmental stresses are involved but not others. Also, these relationships seem to be different depending upon the context, specifically whether it is occurring within a high- or low-income country. Table 8.2 summarizes whether there is a connection between conflict and different types of environmental stress, and whether the relationship changes between high- and low-income countries. It shows, for example, that land degradation leads to greater risk of armed conflict in high-income countries but there is no relationship or connection between land degradation and armed conflict in low-income countries.

In what way do the relationships support or challenge the ideas of Malthus, Marx, and Boserup we introduced at the start of the chapter? What geopolitical structures do you think are relevant? What forms of geopolitical agency do you think are relevant? How do you think these structures and agency interact? How and why would the interaction be different in high- or low-income countries?

Table 8.2 Climate change and the risk of conflict

Environmental stress	High-income countries	Low-income countries
Population growth	Increased risk of conflict	Increased risk of conflict
Population density	Increased risk of conflict	Increased risk of conflict
Land degradation	Increased risk of conflict	No relationship with risk of conflict
Water scarcity	Increased risk of conflict	Increased risk of conflict (but less so than in high-income countries)
Population growth and land degradation	Some risk of conflict	No relationship with risk of conflict
Population growth and water scarcity	No relationship with risk of conflict	Increased risk of conflict

Source: Raleigh and Urdal, 2007

resource that is proximate (such as cropland) is most likely to result in mass rebellions or peasant uprisings that seek to overthrow the existing government and replace it with a new one. However, if the diffuse resource is distant, as is often the case with timber, it is most likely to provoke warlordism – or the use of violence and intimidation to secure de facto control over a region within a state.

Le Billon's (2005) framework forces our attention to the different geographies of civil war and their connection to resources. However, we should not forget that Le Billon situates these intra-state geographies within a context of global politics and the commodification of resources. Oil is the resource that, in today's world, is most commonly seen as a driver of global geopolitical activity, and it is to that topic we now turn.

Oil, empire, and resource wars

The geopolitics of oil is nothing new. Since the end of World War I the recognition of the importance of oil has produced calculations regarding colonialism, the need to maintain friendly governments in oil-rich countries, securing seaways through which oil is transported and (increasingly) pipelines, and claims that military interventions are driven by the interests of oil companies. The invasion of Iraq in 2003 and the support for anti-Colonel Gaddafi rebels in Libya in 2011 have been portrayed as serving, primarily, the demands of oil companies. The immediacy of the geopolitics of oil has been heightened by the concept of "peak oil"; that we have reached the point where accessible oil reserves will diminish as demand grows.

Economic growth in Asia, especially China and India, will increase global demand (Klare, 2009). The increasing profile of Chinese investment in oil production facilities across the globe has provoked some accusations that China is trying to secure reserves and therefore promoting a risk of shortages for other states (Jiang and Sinton, 2011). On the contrary, claims a report by the International Energy Agency (Jiang and Sinton, 2011), rather than being driven by national geopolitical equations, investments by Chinese oil companies are largely independent of national geopolitical and state-based imperatives, but are sound financial investments based on business strategy. Furthermore, rather than creating shortages, these investments have usually increased the global supply of oil that is available through the international market. For example, despite investing heavily in Kazakhstan's oil industry, and the existence of a pipeline into China, some of the oil produced as a result of Chinese investment is sent to China whilst some is sold on the global market.

The discussion of Chinese investment highlights two relevant themes. First, there is the recognition of the importance of a global market that sets the price of oil based upon supply and demand. Second, there is a concentration upon particular territorial geographies, the states investing, buying, and producing oil. The geopolitics of oil is a combination of territorial control and being able to have influence within economic flows of investment and supply. Though there are some critical commentators eager to throw the charge of "empire" at the US for its military actions that are seemingly provoked by the need to access oil reserves, a more careful analysis shows that there is more to the US geopolitical code than territorial control.

Geographer John Morrissey has studied the history of US involvement in the Middle East, the world's primary oil producing region. Since 1945, when President Franklin D. Roosevelt committed the US to backing King Abdul Aziz ibn Saud's regime in Saudi Arabia, successive US administrations have made protecting the flow of oil from the region a priority of their geopolitical code. It has been a non-partisan issue (Morrissey, 2008). However, the situation changed in the 1970s when the US's loyal ally the Shah of Iran was overthrown by Islamists around the same time as the Soviet Union's invasion of Afghanistan. The response by President Jimmy Carter was a reaffirmation of the US geopolitical code toward the region in what became known as the Carter Doctrine:

> Let our position be absolutely clear: An attempt by any outside force to gain control of the Persian Gulf region will be regarded as an assault on the vital interests of the United States of America, and such an assault will be repelled by any means necessary, including military force.
>
> (President Jimmy Carter, State of the Union
> Address, January 20, 1980)

An immediate act of geopolitical agency was the establishment of the Rapid Deployment Joint Task Force (RDJTF). Morrissey (2008) argues that the Carter Doctrine and the RDJTF were not just a function of the 1970s rise in oil prices and the related crises in Iran and Afghanistan, but rather of the broader geopolitical context of the Cold War, including nuclear parity with the Soviet Union and wound-licking after defeat in Vietnam. The Carter Administration had been considering the Middle East as a region in which superiority over the Soviets could be reestablished. The geographical focus of this global calculation became clear when the "area of concern" of the ostensibly global and newly formed RDJTF was established as the Middle East and the Horn of Africa (Morrissey, 2008). The incoming administration of President Ronald Reagan reorganized the RDJTF into a separate regional command, central command or CENTCOM, and in the process promoted the region as the most important focus of the US geopolitical code above Western Europe and Northeast Asia. The regional emphasis of the Carter Doctrine and the establishment of the RDJTF and CENTCOM suggest that, for the US, the geopolitics of oil is about the military control of territory (see Figure 8.5).

However, when the economic motives of CENTCOM are investigated, other geographies become apparent. The presence in the Middle East is to maintain and promote not just US strength but the economic vitality of the *world* (Morrissey, 2008, p. 108). Since 1983 every CENTCOM Commander has gone before the US Congress to affirm the connection between US military presence in the region and the global economic benefit. As General Norman Schwarzkopf put it at the time of the US mission to expel Iraqi forces from Kuwait in 1990:

> the greatest threat to U.S. interests in the area is the spillover of regional conflict which could endanger American lives, threaten U.S. interests in the area or interrupt the flow of oil, thereby requiring the commitment of U.S. combat forces.
>
> (Schwarzkopf, 1990 quoted in Morrissey, 2008, p. 112)

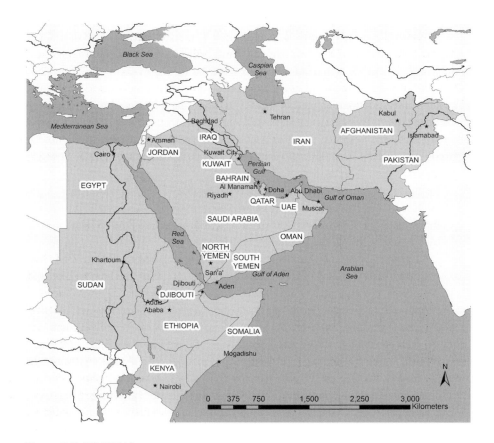

Figure 8.5 CENTCOM.

Or as a subsequent CENTCOM Commander in Chief, General James Binford Peay III, said in 1994, the mission of maintaining "regional stability in the Persian Gulf [was] integral to the political and economic wellbeing of the international community" (Binford Peay, 1995 quoted in Morrissey, 2008, p. 113). These goals continue in the geopolitical practices and representations of the US, and were reiterated during the administration of President George W. Bush in the 2005 National Defense Strategy that noted the importance of a military presence in the region and the need to protect "the integrity of the international economic system" (Morrissey, 2008, p. 113).

The connection between the oil resources of the Middle East region, the global economy, and US geopolitics is reinforced by the analysis of geographer David Harvey (2003). He argues that the connection rests upon the following proposition: "whoever controls the Middle East controls the global oil spigot and whoever controls the global oil spigot can control the global economy, at least for the near future" (2003, p. 19). Military actions such as those in the establishment of CENTCOM and the invasion and occupation of Iraq are seen as means to stave off economic competition from China and the EU by controlling the cost of oil, and the manner in which it is distributed across the globe (2003, p. 25).

The contemporary geopolitics of oil is a complex mixture of global supply, increased demand related to economic growth (and especially the trajectory of India and China), a territorial focus upon military presence in the Middle East, and the flow through trade networks of oil exports. Territorial geopolitics meets the geopolitics of flows and their control, and it is all occurring within a geopolitical context of competition between states. The geopolitics of oil reinforces the themes we have emphasized throughout the book: that making geographies and making politics are necessarily intertwined. This connection is very evident in environmental geopolitics as the physical landscape is transformed by political actions. We conclude the chapter with a case study to highlight how politics remakes places and the physical landscape.

Case study: counter-insurgency turning jungles into forests

Conflict transforms the environment. For example, in the Vietnam War the US used a chemical called Agent Orange to defoliate thousands of acres of forests to hamper the ability of North Vietnamese to hide. Another example is the irradiation of atolls in the Pacific Ocean through their use as nuclear bomb testing, making then uninhabitable and requiring their populations to be relocated. Numerous parts of the world, often areas of natural beauty, are used by the military as shooting or bombing ranges, inhibiting their use by the public and creating vast areas of the planet fenced off in the name of national security (Woodward, 2004). By taking a more detailed look at one particular set of events we can see how security and conflict become driving imperatives for changing the environment physically as well as describing it in a particular way. In other words, geopolitical practice may include transformation of the environment as well as the way the landscape is represented.

In a series of studies of Southeast Asia geographers Nancy Peluso and Peter Vandergeest have shown how insurgency and counter-insurgency have increased state management of parts of the natural environment. Part of this transformation has been a change from the label "jungle" to "forest," or from something that is wild and dangerous to something that is controlled and economically useful.

After World War II, Southeast Asia experienced the geopolitics of decolonization. The Japanese empire had been defeated in the war and the global move toward independent nation-states produced movements forcing the end of British, Dutch, and French colonial involvement in the region. The geopolitical process of decolonization became entwined with the geopolitics of the Cold War, such that the lines and motives between anti-colonial and Communist movements became blurred. The result was a series of conflicts that involved insurgents against colonial authorities and, once independence had been gained, ideological and separatist insurgencies against the new states. Though Korea and Vietnam are perhaps the most remembered of such conflicts, there were also insurgencies and counter-insurgencies in Indonesia, Malaysia, and Thailand. These last three countries, and the period from the 1950s through the 1980s, are the focus of Peluso and Vandergeest's (2011) work.

The process of creating new nation-states and fighting against insurgencies combined to create national forests as controlled and profitable areas. Instead of being jungles that were identified as being beyond the control of the state, and therefore havens for rebels or insurgents, vast areas of these countries became managed national forests. In the process, the institutions and personnel of the state became active and visible in these areas of the national territory and played a role in making the inhabitants national citizens.

Peluso and Vandergeest (2011) identify four connected parts of this process:

1   Forested areas played a key role in insurgencies by providing a haven and strategic territory. These areas then became geopolitical entities.

2   In targeting forested areas they became transformed from "jungles" into "national forests." This was not simply a representational switch. It also involved the establishment of new institutions to manage the forests and the people who lived within them.

3   The state response to jungles/forests as geopolitical entities entailed the "massive spatial reorganizations of populations through resettlement, colonization, and the territorial rezoning of property rights" (Peluso and Vandergeest, 2011, p. 589). This required a process of racial labeling of the segments of the new national populations, producing majorities and minorities. Hence, managing forests as part of insurgency/ counter-insurgency was used as a component of the top-down nationalism we discussed in Chapter 4.

4   Fighting insurgents in "jungles" and then controlling minority populations in "forests" involved a process of militarization of these parts of the environment. Controlling the population went hand-in-hand with economic development. "The mobilization of troops for fighting, patrolling, and other security activities, as well as the military's use of surveillance technologies, were consistent with the needs of forest managers for forest protection and surveillance. The expense of such technologies had previously precluded their extensive development and application for forest surveillance, particularly before the timber industry became an important part of these national resource-extracting economies and before the rise of international conservation" (Peluso and Vandergeest, 2011, p. 590).

This case study does more than illuminate the relationship between conflict and the transformation of the environment. It also is an example of some of the basic concepts and ideas we introduced in Chapter 1: militarization of forested areas, the development of timber industries, and the transformation of "jungle inhabitants" into citizens, even as "minorities" in some cases, are examples of processes of social construction. Specifically, jungles were represented as dangerous and ungoverned spaces but became constructed as important places within the nation-state. Geopolitics, or the way politics and geography construct each other, is seen in the way nation-state building and counter-insurgency transformed "jungles" into national forests, and the nature of the forested areas facilitated militarized government control.

Summary and segue

Human beings are a component of nature, rather than a separate entity that interacts with the environment. Thinking of the environment as a geopolitical structure illustrates how the agency of states and international organizations creates the structure of the global environment and the ways in which that structure changes. Contemporary environmental geopolitics is a tension or contradiction between the recognized need for states and people across the globe to connect and cooperate, while at the same time promoting (in some cases) agency that is based on narrow self-interest. The geopolitics of the environment forces recognition that individuals are parts of broader groups or identities that interact within one world, whether we recognize those connections or not. However, cooperation is difficult to achieve precisely because geopolitical agents are complex and operate within multiple structures. The following concluding chapter focuses upon the complexity of geopolitical structures and agency.

Having read this chapter you will be able to:

- Consider the ways humans interact with the environment
- Understand the concept Anthropocene
- Investigate how a new geopolitics has emerged within the Anthropocene
- Evaluate the causes of resource conflicts
- Consider the interplay between territorial and network geopolitics in resource conflicts

---

Further reading

Dalby, S. (2011) *Security and Environmental Change*, Cambridge: Polity Press.

This short and accessible book provides a summary of the current state of thinking about environmental security and the implications for humanity. It is written by one of the most prominent scholars of environmental geopolitics.

Klare, M. T. (2009) *Rising Powers, Shrinking Planet: The New Geopolitics of Energy*, New York: Holt Paperbacks.

Bringing together the geopolitics of global supply and demand for oil, as well as the politics of control of oil resources, this book extends the themes of our discussion in this chapter. The book also discusses the politics of alternative fuels.

Peet, R., Robbins, P., and Watts, M. (2011) *Global Political Ecology*, New York: Routledge.

A collection of essays that discuss a broad range of resources and environmental issues and make connections between the actions of people and organizations in specific places across the globe with global trends and the power politics of the global economy.

9

MESSY GEOPOLITICS: AGENCY AND MULTIPLE STRUCTURES

In this chapter we will:

- Emphasize the complexity or "messiness" of geopolitics. In other words . . .
- Highlight the interaction of multiple geopolitical structures in creating specific geopolitical contexts
- Focus on the topic of rape as a weapon of war to illustrate the argument
- Use a case study of the conflict over Jammu and Kashmir to exemplify the concepts
- Consider a geopolitical commitment to peace
- Note how complexity or "messiness" is a product of the interaction of structure and agency

This chapter will conclude our introduction to geopolitics by emphasizing complexity, or "messiness." Each of the previous chapters has focused upon a particular set of geopolitical agents and structures; the geopolitical codes of states, or the meta-geography of terrorist networks, for example. However, in Chapter 1 we introduced agents and structures by talking about how they could be seen as nested scales, in other words any geopolitical agent will have to simultaneously negotiate the opportunities and constraints of a number of structures. Furthermore, geopolitical agents have multiple goals – they are not homogeneous, simple, or singular entities. The multiplicity of geopolitical goals is evident in individuals, nations, states, terrorist groups, and any other geopolitical agent. In other words, geopolitical agents juggle a number of identities, some competing and some complementary. Combining the multiplicity of agents' identities and goals with the combination of geopolitical structures indicates that geopolitics is a messy affair.

Who am I, who am I fighting, and why?

The cause of the Palestinians is commonly identified as a nationalist struggle – the desire of a people for their own independent state. Indeed, this was the focus of the case study in Chapter 5. However, to talk of the Palestinians as a homogeneous group is false. On the one hand, the politics of the national group is the product of competing groups with different goals. For one, factional politics is a key part of Palestinian politics. But instead of focusing on formal or party politics, let us begin with the example of an individual Palestinian man, 'Adnan, living in the Rafah refugee camp in June 2001, and his experiences of an Israeli army raid that demolished seventeen houses, the homes of 117 people:

> While the shelling continued, I took my disabled mother, who requires a wheelchair, and told my wife and children to get out of the house. They were all frightened and hysterical. Throughout the neighborhood there were screams of little children, and adults asking, "Where is my son? Where is my brother? Did they get out?" . . . At approximately 5:30 a.m. it ended. The army left the area, and I looked for my wife and children. My sister Hanan told me that my wife, who is pregnant, was on the main road and couldn't stand on her feet out of fear because of the horrible sight of the demolished houses. I went to her and asked what happened. She said that she was bleeding, a result of fear and the running from the house . . . The army also demolished my irrigation pool, the shed with motors and pumps, and a one-hundred-square-meter sheep pen. The pen had six sheep and one of them was killed during the demolition. The bulldozer also uprooted six olive trees that were forty years old.
>
> (*B'TSELEM*, 2002, 17–18; quoted in Falah and Flint, 2004, p. 124)

Who is 'Adnan, or in what ways can we identify him as a geopolitical agent? Father, father-to-be, husband, son, brother, farmer, and current guardian of an olive grove that would be the hope of income for future generations. At the intense moment of the destruction of his home, what are 'Adnan's geopolitical goals? The quote stressed protection of his immediate family, both in the sense of their physical health and of their economic well-being. Family and economics are the structural imperatives in the quote. Of course, they are linked to his plight as a refugee, and so to his membership of a stateless nation. The limited efficacy of his agency must be understood in relation to the coercive power of the Israeli Defense Forces, as well as the meta-geography of global geopolitics; legitimacy is accorded to states, and their citizens have rights that are not possessed by refugees.

The next example is also intended to illustrate that geopolitical agents operate within a number of geopolitical structures, even if they are not conscious of them (Figure 9.1). Darfur is a region of western Sudan, and home to ethnic Africans. The region has been in conflict with the Arab-dominated Sudanese government. Calls have been made to classify the killings in the region as genocide, sponsored by the Sudanese government. The international community has been reluctant to label the killings, as widespread and systematic as they are, genocide because it would produce expectations of intervention. The violence has been committed by Arab militias known by their victims as *janjaweed*, or "devil's on horseback," widely believed to be doing the bidding of the government, though this is denied by officials in Khartoum. The UN investigation of the attacks has highlighted the use of rape as a weapon by the *janjaweed*. The reported incidents are numerous, but to give a sense of the horror one attack in March 2004 involved the rape of 16 girls by 150 soldiers and *janjaweed*. It is alleged that girls as young as 10 have been raped.

The geopolitical implications of rape as a weapon of war are discussed in Box 9.1. Here the topic of multiple geopolitical structures is illustrated by a February 14, 2005 article from the *New York Times* that reported on the difficult future of the babies that are being born as an outcome of systematic rape in Darfur. Fatouma, a 16-year-old mother and rape victim, identifies her baby as a *janjaweed*. "When people see her light skin and her soft hair, they will know she is a *janjaweed*" (Polgreen, 2005, p. 191). For now, the baby is being raised and protected, but the future for both mother and baby is uncertain given the deep cultural taboos regarding rape and the fact that, in the Muslim tradition, identity is passed from father to son. The article continues by quoting the village sheik's thoughts toward the baby, "She will stay with us for now . . . We will treat her like our own. But we will watch carefully when she grows up, to see if she becomes like a *janjaweed*. If she behaves like a *janjaweed*, she cannot stay among us." Ethnic identities of them and us, as well as the position of women in a patriarchal and traditional society, interplay to make the future for women such as Fatouma and her offspring bleak. Fatouma's goals are clear: "One day I hope I will be married . . . I hope I find a husband who will love me and my daughter". The structures of religious and ethnic tradition and honor make the accomplishment of these goals problematic. In the *New York Times* article, another rape victim reports how her husband has abandoned her and her children, one a product of rape, with little hope of remarrying and, hence, facing a lifetime of economic hardship and social isolation.

Figure 9.1 Child soldiers.

Box 9.1 Rape as a weapon of war

Rape is increasingly used as weapon of war: it "routinely serves as a strategic function in war and acts as an integral tool for achieving particular military objectives" (Ramet, 1999, p. 206). Recent and ongoing conflicts in the former Yugoslavia, Rwanda, Darfur/Sudan, Burma/Myanmar, Jammu, and Kashmir have all involved systematic rape. Rape is an effective weapon because it has an impact upon a number of geopolitical structures and, hence, is disruptive in many ways.

For example, Allen's (1996) discussion of rape in the former Yugoslavia points to the ability of systematic rape to change the perception of place; after public rapes of Bosnians and Croats had demonstrated the danger of remaining, people would leave their established homes and leave the vacated town for occupation by Serbs. Rape had changed a place from a traditional site of community to a venue of fear, and so facilitated the brutal redrawing of the ethnic geography of former Yugoslavia.

Rape in warfare is also a means of enforcing pregnancy "and thus poisoning the womb of the enemy" (Crossette, 1998). From this perspective the target of the rapists is at the individual scale of the mother and the offspring. The woman becomes "damaged goods in a patriarchal system that defines woman as man's possession and virgin woman as his most valuable asset" (Allen, 1996, p. 96). As one Rwandan rape victim said, "We are not protected against anything . . . We become crazy. We aggravate people with our problems. We are the living dead" (Human Rights Watch, 1999, p. 73). Rape victims are unable to find husbands and bear other children, and hence become rejected by their families and communities. The target of the rapist in war is also the child in a context in which membership of one ethnic community is vital and children born from rape can be seen, for example, as infusing Serbian blood into other ethnic groups and producing "little Chetniks" or "Serb soldier-heroes" (Allen, 1996, p. 96). Rape destroys the life of the individual and disrupts the identity and cohesion of the community and the ethnic group.

Key to understanding the ability to motivate soldiers to rape as well as the disruption of communities is the notion of patriarchy. The ability to violate and harm women, to see rape as an acceptable form of combat, requires soldiers to be socialized within structures that see the domination and control of women as a norm. The strategic understanding that rape victims will be rejected by their communities and families also rests upon the patriarchal view of women as "property" that cannot be married off or produce wanted children after rape.

Understanding the power of patriarchy is crucial in making sense of the impact of systematic rape, and hence its adoption in war. In a nationalist or ethnic conflict, when it would appear that group identity is the dominant geopolitical factor, a daughter attacked by the enemy group does not receive the sympathy and help of

(Continued)

her own community. Patriarchal values trump communal solidarity. In the former Yugoslavia, for example, women feared that they would be shunned by family and friends (Allen, 1996, p. 70), and the victims' trauma was "exacerbated by cultural taboos associated with rape" (Human Rights Watch, 1999). In Jammu and Kashmir, Pandit and Muslim women who suffered rape in the conflict were taunted by their neighbors (of their own cultural group) and sometimes cast out by their families. After rapes committed by the Indian security forces in 1991, "women had been deserted by their husbands . . . a seventy year old woman had been thrown out by her son . . . [and] girls . . . were teased even by the village men" (Chhachhi, 2002, p. 200). National and community solidarity in the face of conflict took second place to embedded views of the status of women. However, the very rejection of women rape victims by their own communities disrupts societies and cultures, and so is seen as an effective weapon of war.

The final geopolitical structure I will introduce in this discussion is the state. The case of Myanmar/Burma is especially indicative, though certainly not the only case, in which the government is active in promoting its soldiers as rapists. Systematic rape by the army is based upon a patriarchal society, with "many indicators of male predominance and female subordination throughout Burmese society" (Apple, 1998, p. 26). Also, the army is alleged to "recruit" teenagers by kidnap, and one argument is that systematic rape by Burmese soldiers is indicative of the abuse they have suffered themselves (Bernstein and Kean, 1998, p. 3). The Burmese government uses the army in its attempt to dominate minority ethnic groups. Similar to other conflicts, rape in Burma is used to illustrate the power of the state over ethnic minorities, to instill fear, and nullify any plans for rebellion (Women's Organizations from Burma, 2000, p. 27). Furthermore, Burmese soldiers are taught that by impregnating women from ethnic minorities they will be leaving Burmese blood in the villages which will end the rebellion. Perhaps unique to the case of Burma, and indicative of the combined domination of state apparatus and patriarchy, is the belief that rape provides the opportunity for soldiers to give women "pleasure" and so persuade them into a marriage that would diffuse Burmese "blood" and diminish the minority population (Apple, 1998, p. 44).

Rape as a weapon of war is an important topic to discuss because it is illustrative of the manner of fighting in the civil and ethnic wars that are most common today. Theoretically, the issue of rape in warfare reveals that geopolitical agency is often very aware of the multitude of geopolitical structures and their interrelationships. By targeting relatively weak individuals, an army can disrupt communities and cultures. However, such belief in the strategy, and its chances of success, are made possible by existing patriarchal structures that view women in particular and subordinate roles.

The two examples in this section illustrate that the geopolitical experience is immediate and personal. Individual geopolitical agents experience hopes and barriers within family, household, and local structures. The "all-seeing" global visions of academics such as George Modelski not only are abstractions compared to real experiences, but also identify structures seemingly removed from the immediate horror of rape and social rejection, or the destruction of one's home by bulldozers. It is geopolitical structures that help us understand the reasons behind the situations facing 'Adnan and Fatouma, as well as the potential change for the better, however remote. Conflict between the people of Darfur and the Sudanese government, and also between Israel and the Palestinians, is framed within a meta-geography invoking the importance of collective national and ethnic identities, a them-and-us mentality requiring territorialized notions of belonging.

What of Modelski's model? Can we relate the geopolitical structure of cycles of world leadership to the complexity of individual actors? In some cases, I think so. On Monday February 14, 2005, the *New York Times* ran a report in its New York/region section, containing stories that address the local interests and issues of the New York readership. The report was the first in a series entitled "Deployed" and related the experiences of New Yorkers serving in the US armed forces. The byline of the story is Forward Operating Base Speicher, Iraq – an interesting geography for a "local" story. The story tells of the 42nd Infantry Division of the New York National Guard (akin to "reservists" or the Territorial Army in Great Britain) who had been deployed with other National Guard and regular soldiers to be responsible for the security of four Iraqi provinces. The report states that it is the first time a National Guard Division has been fully deployed in combat since the Korean War, and it is also believed to be the first occasion in which a National Guard unit has had command over regular units in a combat situation.

Deployment is not a matter of a division, it is the experiences of individuals – in the case of the National Guard, people who are extracted from civilian jobs in a "local" place and put into combat across the globe. As the *New York Times* points out, "The soldiers here are the men and women who deliver your mail, cook your entrees, answer your customer service calls and patrol your streets. They are truck drivers, students, social workers, youth counselors, cosmetologists, doctors, mechanics, firefighters, general contractors, a pool repairman, a tea salesman, the manager of a coffee shop and a lawyer from Long Island who said he represented 'slumlords'." The article contains the requisite quotes from those who see their service as a matter of pride and national service at a time of war. Major Sal Abano, a 41-year-old chief information officer for an insurance company, in civilian life, is certainly proud. He also makes a geographic connection: "I think it's good for New Jersey to have a presence here." Sentiments echoed by Specialist Dominick Schoonmaker, a 34-year-old mechanic, who relates his service directly to the terrorist attacks of 9/11 and says, "You can't sit on the sidelines anymore."

Others in the Division reflect "an undercurrent of regret and dread evident in many National Guard troops who acknowledge that they found themselves in a situation they neither imagined nor wished for." The brave confession of 43-year-old Staff Sergeant David C. DeMaio is "When I got activated I cried like a baby." Given their recent and central roles as civilians in their communities, the deployment of the 42nd connects places in New York and New Jersey to places in Iraq. Sergeant Major James D. Rodgers is, in civilian life, a letter carrier (postman) and says that his customers have kept in contact with him through e-mail and regular post, "They're very interested in how I'm doing out here," he says.

In Chapter 7, we related the 2003 invasion of Iraq and the subsequent military occupation to Modelski's model of world leadership and, specifically, the need to place combat troops across the globe in the face of challenge to the leader's authority. Although one uses an abstract model to identify long-term trends and structures, the global politics of world leadership is enacted by individuals. These individuals have multiple goals and identities; simultaneously proud defenders of their country and scared family members hoping to return to the tranquility of civilian life. The pride in being a combat soldier stems from the role of the US as world leader and its representation as the global protector of "freedoms" and human rights. This representation is given greater resonance to individuals through narratives of national history.

The final quote from the article is the most significant though. The extended use of the National Guard has made a strong connection between communities in the United States and the combat zone, facilitated by email and websites. The impact on communities and families as men and women fight abroad is made clear, and is able to be communicated from combat zone to home front more quickly than in the past. Structures of family, community, military, nationalism, and global politics are tightly connected. In such a situation, the geopolitical representation of the war is a vital battlefield – if Sergeant Major Rodgers is able to report back to his community on what the combat experience is like, the need for his sacrifice to be seen as vital and moral is enhanced.

Figure 9.2 Returning from war.

Modelski's model is an abstract simplification. The story of the National Guard places "people" within that model. By adding real and particular individuals into the picture the descriptive nature of Modelski's model becomes less dominant, and instead we are forced to ask questions as to how and why actors operate within the structure. Noteworthy is the readiness with which the soldiers in this newspaper story say things which reflect the world leadership role of the United States. In other words, the language used by the US government in its world leadership role is echoed by individuals. Despite the pull of other identities (family, home, career, community, for example), identity and loyalty toward the United States nation-state *in its global role as world leader* are readily and unquestioningly offered. However, the actions of the National Guard members are no means predetermined. Their loyalties to family, for example, in addition to the way their actions are interpreted in their home communities (especially if opinion toward their military mission becomes negative) suggest that the level of their commitment may change. In other words, they are not automatons within a geopolitical structure, but complex agents who make decisions. It is well within the realms of possibility that people choose self and family over military service, for example. Indeed, the purpose of military recruiting campaigns is to ensure that people choose national service, while the modern military is very aware that long overseas deployment puts stress on the family and decreases retention.

Let us compare the geographic connectivity of the National Guard in a contemporary conflict with the experience of citizens drafted to serve in World War II. Fussell (1989, p. 288) notes the way war was reported and the self-censorship involved while writing letters home meant "a slice of actuality was off limits." In a statement which could be applied beyond the single case of World War II, he goes on to argue that war's "full dimensions are inaccessible to the ideological frameworks that we have inherited from the liberal era" (1989, p. 290). For Fussell the meaning and experience of war could never be comprehended except by those who witnessed combat.

Activity

Does the contemporary participation of citizen soldiers in relatively easy contact with the "folks" back home change this situation and break down the barriers between battle experience and home? It has been argued that the digitization of war, including the computerized scenarios played out in the media and the US army's own website, produce a virtual war experience that is intense but quite false (Der Derian, 2001). To consider the implications of the differences between now and World War II, think back to the discussion of Orientalism in Chapter 3 and the way that the casualties suffered by the Iraqis have been removed from our understanding of the war in Iraq. In what way is war experienced "back home" depend upon images and understandings of "our soldiers" compared to the casualties of those they are fighting and the civilian population? Do we think differently of soldiers from our country if they are in intense gun-fights or bombing people from a distance with high-tech weaponry? Do our opinions of war change if we see media coverage of our troops in action? Would public opinion regarding the war in Iraq, for example, change if we heard daily reports of enemy casualties?

The focus on the "messiness" of geopolitics shows that the reality and experience of conflict is very different from the simplicity and singular explanations provided by, say, Modelski's model or definitions of nationalism, etc. However, that does not mean that models and theoretical concepts are unimportant. The role of models and concepts is to simplify in order to understand key structures and processes. The trick is to realize that any given situation is the coming together of a variety of geopolitical agents and structures operating at different scales. In other words, we can try and make sense of the "messiness" by first identifying the multiple structures and agents at work and, second, seeing how they interact in complementary and competing ways. The following case study is of the nationalist conflict in the disputed region of Jammu and Kashmir. As well as providing background to this particular conflict, the case study will attempt to show the interaction between geopolitical agents and structures, namely political parties, state officials, nationalist groups, and individuals.

Case study: persistent conflict in Jammu and Kashmir

Timeline

1752: Afghan warlord Ahmed Shah Durrani conquers Kashmir after the collapse of Mughal power.

1819: Owing to the bloody nature of Afghan rule, the Muslim majority surrenders all of their land to Punjabi Sikh King Ranjit Singh so that he may take over, beginning twenty-seven years of Sikh rule.

1820: Ranjit Singh makes Gulab Singh Raja of the state of Jammu. (Starting here and continuing until the British takeover, Gulab Singh begins to build a small empire, expanding his rule into the northwest regions now known as the Northern Territories.) At the same time, Ranjit Singh gives the province of Poonch to Gulab Singh's brother, Dhyan Singh.

1846: Battle of Sabraon, British capture Lahore by defeating the successor of Ranjit Singh. In the resulting Treaty of Amritsar the British award Gulab Singh with Kashmir (he did not send troops to help resist the British) and the title of Maharaja, and the British retain "supreme control of the Valley" (Malik, 2002, p. 19).

1847: Gulab Singh dies and is succeeded by his son, Ranbir Singh.

1858: Beginning of the British Raj (occupation).

1885: Ranbir Singh dies, and is succeeded by his son, Pratap Singh.

1906: All-India Muslim League founded to protect the rights of Muslims in British India.

1925: Pratap Singh dies with no heir. Hari Singh becomes Maharaja of Kashmir and Jammu. He is appointed by the British.

1931: Culmination of Kashmiri Muslim grievances:
- April 29, Khutba (an important Islamic sermon) is banned at a mosque in Jammu.

- June 25, a Pathan cook named Adbul Qadir makes an "impromptu, highly inflammatory speech condemning Hindus in general and Hari Singh's rule in particular" (Malik, 2002, p. 34). He is immediately arrested.
- July 4, Hindu police official, during an incident with a Muslim police constable, tears up a copy of the Koran.
- July 13, 7,000 gather for trial of Qadir. Police open fire, killing twenty-one. Anti-Hindu riots occur all over Srinagar. This day is understood to be a turning point in relations between the Maharaja and his people.

1932: Formation of the All-India Jammu and Kashmir Muslim Conference Party, with Sheik Mohammad Abdullah as president of the new party. Initially this group unites all of the Muslims, but divisions start occurring soon afterward.

1935–6: Poonch is integrated as part of Jammu and Kashmir (result of a lawsuit in the British Indian courts).

1939: Muslim Conference Party, headed by Sheik Abdullah, recognizes the need to secularize; change from Hindu vs. Muslim mindset to lower class vs. elite; name changed to the National Conference Party.

1941: Official re-emergence of the Muslim Conference Party. The most notable figure is Yusuf Shah who has had long-term ideological differences with Sheik Abdullah. Generally speaking, the National Conference is supported in Kashmir and the Muslim Conference in Jammu.

1942: Congress's "Quit India" campaign begins, with the goal of ending British rule of India.

1943/44: Numerous attempts to unite the Muslim Conference Party and the National Conference Party fail.

1944: The Muslim Conference Party's "New Kashmir" campaign emphasizes the desire to achieve rights for all, especially women.

1946: Yet another failed attempt to unite the two parties, motivated by the National Conference owing to its declining popularity. In May, "Quit Kashmir" campaign against Dogra rule, specifically Hari Singh, calls the Treaty of Amritsar illegitimate.

1947: June 3, British announce plan to partition India. August 14/15, ending their rule, British create the two separate independent states, Islamic Republic of Pakistan and India (the Radcliffe Boundary Commission is in charge of setting boundaries).

> At the time of partition the views of the people concerning Jammu and Kashmir fell into three general categories: Hindus (geographically concentrated in Jammu) wished continued rule of the Maharaja. The Muslim Conference members wished to be a part of an Islamic state (either Pakistan or independent) and the National Conference wished to join the secular Indian state. (Muslims were the majority group in the Kashmir Valley and a large amount of Muslims were also in Jammu.)
>
> (Malik, 2002, p. 64)

Mass killings (based on ethnicity and religion) and displacements were occurring at this time. Hindus and Sikhs moved eastward and Muslims were migrating westward.

During the British occupation there were areas that were formerly controlled by the British and areas, such as Jammu and Kashmir, where power was given, by the British, to another leader. At the time of partition, it was assumed that the latter would join either India or Pakistan, based on both geographic location and characteristics of the population. For most of the provinces this decision was clear for either one or both of the reasons listed above, whereas Kashmir and Jammu lay in between the two states, had a majority Muslim population, and were being ruled by a Hindu.

(Malik, 2002, p. 63)

- October 12, statement issued by spokesman for Hari Singh expresses the wish to remain independent and neutral – "the Switzerland of the east" (Malik, 2002, p. 64).
- An uprising, beginning in Poonch, leads to the declaration of an independent Azad Kashmir by its Muslim majority.
- October 26, faced with incoming Pakistani tribal troops, Hari Singh is forced to sign the Instrument of Accession to India. India sends troops to secure the area.

1948:
- March 5, interim government formed with Sheik Abdullah as Prime Minister (at this point, Hari Singh still holds title of Maharaja but with little to no power).
- May, Indo-Pakistan war begins, when Pakistan sends its official troops into Kashmir.

1949:
- January 1, with UN intervention (UNCIP = UN Commission on India and Pakistan) a ceasefire stops the war from spreading into the rest of India and Pakistan.
- January 27, the official ceasefire line is declared and remains until 1965.
- The region is now separated into three different administrative parts: the Northern Areas (controlled by Pakistan), Azad Kashmir (independent in theory), and the rest which was controlled by Indian troops.

1952: Dogras' hereditary position is officially abolished. Relations with India through this entire period are ambiguous because accession has still not been ratified. June 24, Delhi Agreement – Jammu and Kashmir are part of India, but with a higher level of autonomy (but by the following year, Sheik Abdullah is involved in conversations about sovereignty for Jammu and Kashmir with US and UN officials). Hindus oppose the Delhi Agreement because it means that they will not be protected by New Delhi from Kashmiri Muslim rule.

1953:
- August 9, New Delhi arrests and replaces Abdullah. Bakshi Ghulam Moham-med is sworn in as new Prime Minister. Protests occur during the next few weeks. Bakshi has little support from the people, needs New Delhi to keep him in power, but Kashmir experiences a period of stability nonetheless.
- October 5, legal framework is laid for formal accession to India and increased power of New Delhi in Kashmiri affairs.

1954: Pakistan signs a military aid agreement with the United States.

1955:
- Soviet leaders visit India, making a trip to Srinagar.
- Plebiscite Front (aka Action Committee) emerges, an opposition group formed by Mirza Afgal Beg and supported by Sheik Abdullah (anti- Bakshi and New Dehli). This group's goals are less centered around autonomy and more geared toward proving the Instrument of Accession invalid owing to the fact that Jammu and Kashmir are only to become part of India after a popular referendum that never occurs.

1956–7: Following years of friendly Indo-China relations, China begins to build a military highway in disputed territory, Aksai China.

1957: January 26, new constitution of Jammu and Kashmir takes effect reaffirming the accession to India. USSR exercises its UN Security Council veto for the first of what will become many times during a discussion of Kashmir initiated by Pakistan.

1959: India sends border patrols into the area under dispute with China.

1962: Small fights break out between India and China beginning the border war (47,000 square miles of disputed land).
- October 4, Bakshi resigns and Revenue Minister Khwaja Shamsuddin is sworn in.
- October 10–November 20, significant fighting between Indian and Chinese forces. The Chinese People's Liberation Army is well prepared for the fighting in the Himalayas – they are in warm, padded uniforms and have previously fought Tibet in the same climate (14,000–16,000 feet altitude). The Indian army has a small budget and ill-prepared troops.
- November 21, China, after accomplishing all goals of land attainment, declares a ceasefire. "Following the ceasefire, China kept most of her claim in Aksai China but gave India virtually all of India's claim in the North East Frontier Agency – about 70% of the disputed land!" (Calvin, 1984). According to official Indian reports the number of Indian casualties is 1,383 troops killed, 3,968 captured, and 1,696 missing. The data are never released for Chinese casualties.

1964: The Action Committee splits – the Awami Action Committee forms.
- May 23, Abdullah travels to Pakistan for negotiations (after being jailed by the Indian government numerous times he becomes a hero in Pakistan).
- May 27, Prime Minister Nehru dies; relations between New Delhi and Abdullah decline rapidly; negotiations with Pakistan fail.

1965: Indo-Pakistan war.
- July, Pakistan begins sending troops, anticipating support from Kashmiris.
- September 6, India's counter-attack crosses the border into Pakistani Punjab.
- September 23, UN-mediated ceasefire. By the time of the ceasefire, Pakistan has suffered approximately 3,800 casualties while India has suffered approximately 3,000 (Indian Express Group, 2001). National Conference Party changes name to the Pradesh Congress Party (extension of New Delhi government).

1966:   January 10, Tashkent Declaration, result of Russian-mediated peace talks. India and Pakistan move back to pre-war borders, repatriate POWs, and re-establish diplomatic relations.

1971:   Indo-Pakistan war. Originally beginning with a civil war in East Pakistan, which becomes Bangladesh by the end of the war; seen as the liberation of Bangladesh from the Indian perspective. In Kashmir, the Plebiscite Front is banned. Abdullah is externed from the state. July 2, the Simla Pact is signed by both sides. They agree to respect the line of control until further resolutions are made.

1975:   Accord between Abdullah and Indira Gandhi, Prime Minister of India. India sees it as firming the union. Abdullah sees it as protecting Kashmir's special status. He returns to power.

1981:   Farooq Abdullah, the sheik's son, takes over office.

1982:   Sheik Abdullah dies.

1984:   Farooq is dismissed in a "drawing room dismissal" engineered by Indira Gandhi. Protest ensues. Farooq is replaced by G. M. Shah, who is an unpopular ruler.

1986:   The government of Rajiv Gandhi (India's new Prime Minister) reinstates Farooq as chief minister – less popular now in Kashmir because of his collaboration with India.

1987:   Insurgency in Kashmir gains momentum from this time on. Farooq blames unemployment, especially of the educated, with about 40,000–50,000 unemployed graduates. Others point to a rigged election forcing a resort to armed struggle. India responds by intensifying its security actions in the region.

1990:   Central rule imposed on Jammu and Kashmir (with dissolving of the Jammu and Kashmir State Assembly). Between 1989 and 1990, 140,000 Hindus leave the Kashmir valley – they go to refugee camps in Jammu.

1992:   Operation Tiger (followed by Operation Shiva) carried out by Indian Security Forces. These security operations have led to allegations of widespread killings and other atrocities.

1998:   Both India and Pakistan begin nuclear testing.

1999:   Indian and Pakistani militaries clash in Kargil.

2001:   December, the Prevention of Terrorism Bill (POTB) is passed: "a repressive piece of legislation that could be used to justify considerable human rights abuses by the government of India, especially in Kashmir, where India is fighting a counterinsurgency war" (Podur, 2002). A terrorist attack on the Indian Parliament in December leads to build up of Pakistani and Indian troops on the border. Tensions are defused, and troops withdrawn, after months of diplomacy.

2002–8: There are numerous attacks on Hindus in the area.

2008–9: Large anti-Indian protests, regarding land transfers related to a Hindu religious site, result in a security response by Indian forces that leads to casualties and increases tensions.

Accurate, reliable information concerning the number of casualties since the beginning of armed conflict in Kashmir is impossible to obtain. Official handouts give the following information for 1990 to 1999: 9,123 members of armed opposition groups; 6,673 victims of armed opposition groups; 2,477 civilians at the hands of Indian security forces and

Figure 9.3 Historical roots of conflict in Kashmir.

1,593 security personnel. However, the Institute of Kashmir Studies, a research center, has estimated a total of 40,000–50,000 deaths since 1989/90 (all information taken from the 1999 report from Amnesty International, "India," pp. 8–9). Since 2001, tensions between India and Pakistan have waxed and waned. In early 2005, there were signs of cross-boundary cooperation that may be interpreted as peaceful overtures, and an earthquake in the region in October 2005 resulted in promises of further such cooperation. However, the situation is delicately poised. Indian politicians are keen to accuse Pakistan of sponsoring terrorist attacks in India (including the brazen attack on Mumbai in November 2008) and the public are quick to claim Pakistani sponsorship of the Kashmiri militants.

Geopolitical agency in Jammu and Kashmir

The timeline emphasizes the actions of the Indian and Pakistani governments, and different national groups. If we explore the viewpoints of some of the major geopolitical agents we will see not only the major points of contention, but also how different geopolitical structures combine to provide a context for agency. Indeed, the purpose of this case study is to emphasize how different geopolitical structures and agents interact. The goal is to show the complexity of geopolitical conflicts. The perspective from the Indian government has been consistent, identifying the violence in Kashmir as "an internal affair of India" (Srivastava, 2001, p. 95). At the beginning an attempt was made to secure Kashmir as a part of the Indian state through Nehru's relationship with Sheik Abdullah and the National Conference in the hope of gaining support from all Kashmiri Muslims. With hindsight, we can see this was a major misjudgment of popular opinion. Consequently, India has had to use violence in its project to maintain Kashmir within the boundaries of the Indian state. In the language we introduced in Chapter 4, the rhetoric of the Indian government has framed the conflict as the maintenance of the Indian state in the face of what they classify as insurgency.

However, it is wrong to see India's policy as singular or uncontested. Different Indian political parties have addressed Kashmir on their platforms. The following political positions reflect the stance of the parties in the 2004 Indian elections. The Bharatiya Janata Party (BJP) or Hindu Nationalist Party spoke of facilitating dialogue with Pakistan – the pressure on the party was to portray itself as being able to pursue a national agenda in a multi-cultural society. Especially, the party spoke of secularization in order that it could claim the ability to work with Muslims in Kashmir and across India and so integrate Kashmir as a part of India (Upadhyay, 2000). The religious and ethnic identities that had provided the BJP with its electoral success had to be negotiated by the party, in an attempt to make peace across national and religious lines.

Another major Indian party, the Indian National Congress, emphasized security issues, especially the threat of terrorism, in its political campaign. The party's website juxtaposed the meta-geography of terrorist networks with a notion of a harmonious multi-national state: "Indian National Congress will forcefully resist all attempts at using the issue of cross-border terrorism to polarise our society on religious and communal lines" (Sankalp, 2003). Defense of state boundaries was used by the party to make a claim for broad national support.

The conflict in Kashmir extended beyond the immediate region. The tension between Hindus and Muslims was focused upon geographic locations religiously significant to both groups. In 2002, fifty-three Hindus were killed in a terrorist attack on a train that was returning them from a religious voyage to Ayodhya. They had started to plan the erection of a Hindu temple at this site, which is of importance to both Muslims and Hindus (Lineback, 2002, p. 1). The bloodiest example of this type of conflict in India occurred in Mumbai (Bombay). Here riots broke out after Hindus destroyed a Muslim temple that they believed to be built upon the birthplace of their god, Rama. Eight hundred people died during these riots (2002, p. 1). Conflicts throughout India continue to occur. It is primarily a Hindu country but has the second highest number of Muslims (136 million, 14 percent of the population) after Indonesia (2002). The nature of these conflicts illustrates the reflexive consideration of religion; in other words, religious loyalties are entwined with nationalist struggles. Religious identity reinforces conflict over the structure of the nation-state, and simultaneously religious organizations and beliefs are reinforced within a context of nationalist struggle (Stump, 2005).

The official position of Pakistan takes a different approach to the conflict by trying to make a moral argument of national self-determination. On the official website of the Islamic Republic of Pakistan, amongst nine topics (including "Government," "Country Profile," "Economy") is the topic "Kashmir." In the "FAQ's" portion on Kashmir, the Pakistani government states through the very first answer that Kashmir is different from other territorial disputes, in that

> the territory involved is a whole country . . . Here the matter is not one of placing a few hundred square miles on one side or the other of an international frontier and thus settling a boundary conflict. It is a matter of the disposition of a country through the same process by which the two contestants, the Indian Union and Pakistan themselves emerged as independent states – the process of establishing sovereignties on the basis of popular consent.
>
> (Islamic Republic of Pakistan, 2004)

Furthermore the website continues to explain that neither the Maharaja's signature on the Instrument of Accession nor the support of Sheik Abdullah and the National Conference legitimizes the accession to India. This is based on the argument that these acts were not done with the consent of the Kashmiri people. The government of Pakistan's argument resorts to the ideology of nationalism, that the will of a nation demands a state.

The conflict is not just about one state versus another, or even a singular nationalist claim, though. Religious identity, ethnicity, age, and gender are all important structures that combine in different ways. People in the Muslim community have experienced severe treatment from the ever-present Indian security forces. "What unites disparate ideologies and programmes as well as ordinary people is a common enemy – the security forces" (Women's Initiative, 2002, p. 90). Owing to mistreatment, a feeling of favoritism of the Indian government toward Hindus, and unfair elections, militant groups (known as *tanzeems*) have arisen.

"The main political division among Kashmir Muslims now is between those wishing to accede to Pakistan and those wanting an independent state" (Malik, 2002, p. 357). The similarity among all of the Muslims, however, is the desire to be free from India. *Tanzeems* are responsible for murders, rapes, and kidnappings of both Hindus and fellow Muslims. Because *tanzeems* are plentiful and uncoordinated, rivalries result that spur violence between groups. Fundamentalist groups also attack fellow Muslims who act in a way that violates their ideologies. For example, a teenager reports that his father was murdered because he consumed alcohol: "The Hizbul Mujahidden had warned him about drinking but when he didn't care they killed him" (Chhachhi, 2002, p. 201).

In 1987, following elections that were thought to have been fixed (given the very poor performance of the Muslim United Front in the elections), the youth of Kashmir began to protest and many were arrested. Disaffected youth, who believed they were persecuted for their religious beliefs and ethnicity, were sought by the Inter-Services Intelligence

(ISI – the Pakistani Intelligence Service) who promised "arms and training to these 'boys' to launch armed struggles against India" (Santhanam *et al.*, 2003). These recruits came mostly from the Islamic Students League (ISL). In the mid-1990s, due to the enormous amount of casualties within the *tanzeems*, there was a "drying up of young Kashmiri recruits . . . . School dropouts and rowdy elements began to dominate . . . . Rape became common in the Valley while innocent civilians were murdered on the suspicion that they were 'informers'" (Santhanam *et al.*, 2003, p. 28). According to a four-member all-woman team who set out to assess the situation in 1994: "Many people reported the recruitment of thousands of Kashmiri youth from poor families . . . Someone remarked, 'The sons of the rich in India and Pakistan go to America to study, for better opportunities. Our boys go out to learn how to use the gun. The power brokers are not interested in stopping the war, their children are not being sacrificed'" (Women's Initiative, 2002, pp. 89–90). Lack of opportunity for young men, oppression by Indian security forces, a willing sponsor, and nationalist and religious ideology combined to fuel the *tanzeems*.

The *tanzeem* itself could become the most meaningful geopolitical structure, promoting disputes and violence between groups despite claims of a common cause. Brief description of four *tanzeems* shows the mixture of shared and divergent goals. *Hizbul Mujahideen* (HUM) emerged as an important *tanzeem*, headquartered in Srinagar. It is sponsored by the Pakistan government, ISI, and Jamaat-e-Islami, a political party in Pakistan. The objectives of the HUM are to secede from India via armed combat and to merge with Pakistan. In 1990 Jamaat-e-Islami and the ISI took control of Hizbul. Now Hizbul Mujahideen is considered the militant wing of JKJEI. The *Jamu and Kashmir Jamaat-e-Islami* (JKJEI) is closely linked with the JEI in Pakistan. This group was more politically oriented and won seats in the 1987 State Assembly elections (Santhanam *et al.*, 2003, p. 154). This *tanzeem* obviously shares the same objectives as the HUM, but seeks different means. The *Jammu and Kashmir Liberation Front* (JKLF) differs from HUM and JKJEI in its goal of an independent, united Jammu and Kashmir (including Pakistani-occupied Kashmir and the Northern Territories). This group formed in 1988 (when the ISI was easily recruiting angered students and creating many new *tanzeems*) and is headquartered in Srinagar. Finally, the *Jammu and Kashmir People's Conference* (JKPC) is less radical in nature, with the objective of greater autonomy for the state of Jammu and Kashmir under the Indian Constitution. Two points should be taken from the diversity of the *tanzeems*. First, the protagonists in a conflict are rarely unified, and so it is wrong to view a particular cause or issue as singular. Second, the variety of geopolitical structures produced different goals and identities that were mutually reinforcing.

The creation of ethnic difference is also evident in this dispute. A conflict over the location of an international boundary fermented a conflict in which group identity became significant. It is estimated that about 400,000 Kashmiri Pandits (a sect of Hindus with ancestral ties to Kashmir) were forced from their homes between 1989 and 1991. The fears of a Pandit doctor facing a crowd of Hindus outside her house illustrates how cultural conflict was created over time: "Many of the young men in the crowd were boys I had delivered at the hospital! And here they were now shouting for my blood" (Raina, 2002, p. 179). The status of Pandits has changed too, as they have been forced to become refugees: "While the 'refugees' were earlier welcomed and given assistance, local people have now begun blaming them for being the cause of all problems, ranging from typical urban infrastructure shortages of water and transport, to unemployment . . . increased state violence, militant attacks, sexual harassment, etc." (Dewan, 2002, p. 154).

Prior to the recent violence, Pandits and Muslims lived side by side without any problems. One woman recalls that "before the Kashmir issue [her] friends from that region were just Kashmiris; they were not seen as Muslims or Pandits" (Dewan, 2002, p. 149). Now the situation is quite different and Pandits' wishes for the fate of Kashmir differ greatly from those of the Kashmiri Muslims. "They want their own exclusive 'state' within the Valley – Panun Kashmir. This would be a region or state within India, autonomous both from central government and Kashmiri Muslim control" (Malik, 2002, p. 358). In other words, as conflict creates group identity and ethnic violence, the desire for a state for one's own group is seen to be imperative, and the geopolitical structure of a world of nation-states is reinforced.

To end our discussion of this conflict, a consideration of gender illuminates overarching or dominant geopolitical structures, as well as the cracks in their foundations. For the most part, the suffering endured by Muslim women on an individual level in the conflict is practically identical to the situation facing Pandit women who are normally seen to be on the other "side." The common threat of rape (see Box 9.1) illustrates how the structure of patriarchy transcends nationalist and religious conflicts. The perspective of women is also able to stress comprehension of shared values and seek compromise and fusion over conflict and hierarchy. The sentiment of most women is for peace based upon shared experiences. As one Pandit woman said:

> It was after years that we had all gotten together at a marriage – all of us women – Pandit, Muslim, Sardarnis. It was almost like the old days . . . We laughed and danced late into the night. Then, as we prepared to go to sleep, I heard some of the Muslim women whispering among themselves in the next room: "It's been such a lovely evening. It is true, isn't it, that a garden is only a garden of any worth when there are many kinds of flowers gracing it."
>
> (Chhachhi, 2002, p. 207)

However, not all women are united by feminist beliefs that negate the geopolitics of nationalism. A minority of women in the region see their primary role to be within nationalist movements. For example, *Khawateen Markaaz*, originally an organization that carried out social work for Muslims in Kashmir, joined the Azaadi Movement in 1990. This group wishes for an independent Kashmir and believes: "Kashmir is occupied by both India and Pakistan. We are Kashmiri women. We are committed to independent Kashmir. We respect all religions. We are not fundamentalists. People of all religions will live side by side. Kashmiri pandits should come back here, this is their motherland" (Women's Initiative, 2002, p. 86). On the other hand, *Dukhtaran-e-Milat*, begun in 1980, wishes that Kashmir become part of Pakistan. The movement uses the terms Hindustan and *jihad* and its leader says that "if the men make a pact with Hindustan, women of *Dukhtaran-e-Milat* will pick up the gun even against our own men if need be" (Women's Initiative, 2002, p. 87). Clearly, the imperatives of nationalism are more important than submitting to traditional gender roles here. Ironically though, the motivation is far from progressive, as men are to be challenged only if they are seen to be nationalist appeasers.

Many women, the majority who are not themselves a part of militant activity, accept the supporting role to men in their lives who join *tanzeems*. In other words, structures of patriarchy implicate women in the conflict through their subordinate relationship to husbands. Two women from Bandipora express their acceptance of family members taking part in militant activity: "I knew my husband was a militant. I knew that some day he would be killed. I grieve, but I do not complain" (Chhachhi, 2002, p. 194). Regarding her son another woman said, "The child of a freedom fighter will be a freedom fighter" (2002, p. 194).

The case study of Kashmir has emphasized the diversity of geopolitical agents within a particular conflict and the intersection of a number of geopolitical structures. In our framework we note that geopolitical agents have opportunities and constraints set by geopolitical structures. If agents and structures are both multiple, then the choices made by geopolitical agents and the identities that come to the fore will be complex or "messy". However, the "messiness" is a function of the many geopolitical structures that interact to form particular geopolitical contexts. Moreover, geopolitical agents have, well, agency, or the ability to make choices – such as the Kashmiri women who either rejected the language of nationalism or adopted it. By emphasizing "messiness" in this final chapter, two things should stand out. First, no geopolitical conflict is simple – there are divisions among the antagonists, or many different struggles (gender, race, religion, etc.) are in play within what is often reported as a "one-issue" situation. Second, you can understand the complexity by identifying the different structures that are operating, and noting the way they intersect. As a result, an attempt can be made to identify and understand the options (or lack of them) available to the different agents.

If you are in a class or other group setting you could do this project with someone else and explore the same conflict using different media sources. Not only will this help you in identifying more structures and types of agency, but you may also consider how different media outlets emphasize different structures and types of agency over others. For example, were political parties and state ministries or departments emphasized in one source while protest groups, women's groups, and other social groups were emphasized in another?

Activity

Find a news magazine such as *The Economist*, *Atlantic Monthly*, *The New Yorker*, *Time*, or the color supplements of the Sunday newspapers. These magazines usually carry longer stories on current conflicts than the daily newspapers, stories that include interviews with the participants and victims. Explore an article of your choice and use the interviews and descriptions of the participant's circumstances to identify different structures and how they interact. Do the interviews show divisions within particular groups or agents, such as political parties, ethnic groups, etc.? In other words, does the article exemplify how geopolitical agents are not singular?

Messiness, structure, and peace

The pursuit of peace is often represented as the goal of geopolitical agents. States consistently represent their actions, even when they take the form of war, as attempts to create peace. But, as we discussed in Chapter 6, peace can be defined in different ways, and the form of definition is related to the identification of actors and structures. The idea of negative peace, or absence of violence, can lead to a focus upon states as the only meaningful geopolitical actors (Galtung, 1964). States can agree to end wars, or a strong and victorious state can impose a peace on weaker states. Negotiations between states lead to treaties that impose conditions and behaviors that result in a lack of war, or a condition we call "peace." However, the lack of overt violence does not necessarily mean a just and sustainable political situation; meaning that we should be aware of the false dichotomy between peace and war (Kirsch and Flint, 2011a). Indeed, a negative peace often requires the construction of spaces and places in which either the power relations that led to war are continued, or new ones are put in place. Negative peace is then another form of geopolitics, the intersection of power and geography.

An awareness of the problems of negative peace has resulted in a call by geographers to consider what peace is and what it means to study and practice geographies of peace (Megoran, 2011; Williams and McConnell, forthcoming). Megoran (2011) argues for an engagement with the idea of positive peace that requires the integration of society (Galtung, 1964) and engages a range of scales, from the individual body and mind to the global. Positive peace identifies peace as a process rather than a situation or an outcome, a commitment that requires constant engagement and evaluation of power relations and their implications. The process is one in which multiple power relations are involved, and hence multiple forms of agency and structures. Peace is, then, something that cannot be left to states' elites and formal treaties, but requires social groups to constantly engage matters of race, economy, gender, and the environment.

Megoran (2011) challenges geographers not simply to study peace (rather than being focused upon war) but to be *committed* to peace. Particular research agendas, such as critiquing the "peaceful" actions of states and investigating ways in which border disputes may be resolved, are one form of engagement. Another fruitful approach, and one that ensures consideration of a variety of structures and forms of agency, is the increasing attention being paid to everyday peace – or the way in which people create institutions to maintain social harmony – for example Williams' (2007) analysis of Hindu–Muslim relations in the Indian city of Varanasi. Another important topic of study is peace movements and activism, discussed as a form of network geopolitics in Chapter 6.

However, Megoran (2011, p. 8) argues that geographers need to be committed to the construction of a "pacific geopolitics" defined as "the study of how ways of thinking geographically about world politics can promote peaceful and mutually enriching human coexistence." This requires not just studying those who practice forms of feminist geopolitics and antipolitics, but a commitment by geographers to participating in peace activities and practicing forms of nonviolence. The main way academic geographers can do this is through their teaching, including student-based public-engagement projects. In addition, geographers can use the Internet and participate in social movements to express a public and collective engagement with peace (Megoran, 2011).

Whether it is through traditional forms of research and teaching, or through forms of public engagement and activism (Koopman, forthcoming), the pursuit of peace is a noble commitment for a globalized world, facing nuclear proliferation, terrorism, and environmental degradation. The content of this book shows that a commitment to peace as a process is one that goes hand in hand with an understanding of (pacific) geopolitics as a messy interaction of multiple agents and structures.

Conclusion and prologue

A book such as this has no definitive conclusion. The book's task is to let the reader initiate inquiry into geopolitics and not to provide things that are "known." The case studies are included to provide background to what have proved to be persistent conflicts that could intensify and expand. Knowledge of these actual conflicts is necessary to understand contemporary geopolitics in two senses: the basic "what is happening/where is Chechnya?" sense and as a way to exemplify the manner in which geopolitical structures and agents interact. In the first sense the case studies provide a stepping stone toward a knowledge that will steadily expand as you continue to explore and engage current affairs. In the second sense, the case studies are my attempt to talk you through some actual conflicts with reference to the framework of structures and agents – they are an exercise that I hope will facilitate your ability to analyze future geopolitical situations.

If I have one goal with this book it is to make you informed and active participants in geopolitics. In the most everyday sense, I hope that working through this book allows you to critique what you see and hear in the media. When "experts" are put in front of the cameras or framed on the opinion pages do not be in awe. Instead, use the perspective and knowledge you have gained form this book to question their assumptions and approach to the conflict. Reflect on the limited range of questions that are asked. Wonder what someone from another national, gender, class, racial, religious, or political perspective would say instead. To do this, the first thing is to tease out *all* the geopolitical structures and agents that are involved in the conflict and, hence, be aware of what the expert is *not* discussing. The next step is to construct a fuller picture than the expert will deliver by integrating the role of the excluded agents and structures.

My other intention in writing this book is to act as a guide to participating in geopolitics; but I am aware that this is a pretentious claim, so please let me qualify the statement. I hope that one of the lessons from this book is clear: we are all geopoliticians; we participate on a daily basis. We recreate our own national and state structures by simple acts of reading a "national" newspaper that is organized to talk about "them" in the international section as opposed to "us" in the politics, sports, and weather sections (Billig, 1995). We carry around images of other countries and conflicts that are based upon popular representations of geopolitics, which in turn influence our approval of or opposition to foreign policy. Being aware of the structures of global inter-state interaction, and nationalism, may, at the very least, allow for more reflection when one is asked to act in the name of the "common sense" that such structures inspire – a common sense that feminists will be eager to point out revolves around hierarchy, difference, and violent competition. What are the structures and notions of "normal" behavior underlying "Dulce et Decorum Est Pro Patria . . . "?

For many, participation in geopolitics is much more than the passive reconstruction of structures that are remote and somewhat intangible. Career paths may well lead to direct involvement: teaching in the United States, I am responsible for the education of many young adults who have already begun serving in the armed forces or wish to pursue careers in intelligence agencies or as part of the Department of Homeland Security. For many of the current generation of university students, political awareness was initiated on September 11, 2001. Their sense of geopolitics is very much molded by the language of the War on Terror.

Participation in geopolitics is also a matter of questioning and challenging the "common sense" assumptions generated by the geopolitical structures in general (difference, conflict, etc.) as well as by the representations and actions of key geopolitical agents, the US and British governments for example. Protest, dissent, questioning are also evident amongst the students I teach – disaffection with both the persistent structures as well as specific government actions are also common viewpoints that produce their own actions. The commitment to peace that Megoran (2011) asks for is certainly an option I hope we all consider.

I am in no position to be judgmental about the geopolitical actions that others take. The message that I want to end with is that agency is constrained and enabled by structures. You have choices within structures – knowing the structure makes for a more informed strategy – whether that is within a family, neighborhood, business, social movement, or state. The same awareness may be applied when interpreting current events. The decisions made by the governments of Iran and North Korea, for example, may be portrayed as irrational and unnecessarily aggressive by Western governments. But it is your task to see them as agency within structural settings of global, regional, and inter-state politics. As Robert McNamara advised in the excellent documentary *Fog of War*, Lesson Number One is to empathize with your enemy. Knowing the structural context of other geopolitical agents is a means to knowing their fears, concerns, and goals. Such knowledge of geopolitics is an avenue to empathy and understanding that will, I hope, be a pathway to a more peaceful world.

To finish, one may be more poetic in considering structures and agents:

> The world is big. Some people are unable to comprehend that simple fact. They want the world on their own terms, its peoples just like them and their friends, its places like the manicured little patch on which they live. But this is a foolish and blind wish. Diversity is not an abnormality but the very reality of our planet. The human world manifests the same reality and will not seek our permission to celebrate itself in the magnificence of its endless varieties. Civility is a sensible attribute in this kind of world we have; narrowness of heart and mind is not.
>
> (Chinua Achebe, Bates College commencement address,
> May 27, 1996)

Further reading

The readings listed at the end of Chapter 1 as more detailed and sophisticated investigations of geopolitics should be reviewed. They will provide different interpretations and topical concentrations that will be accessible after reading this book.

Allen, B. (1996) *Rape Warfare: The Hidden Genocide in Bosnia-Herzegovina and Croatia*, Minneapolis: University of Minnesota Press.

A discussion of the role and meaning of rape in warfare, with a detailed case study.

Bose, S. (2003) *Kashmir: Roots of Conflict, Paths to Peace*, Cambridge, MA: Harvard University Press.

Provides an understanding of a long-running conflict that has broader regional implications.

Stump, R. W. (2005) "Religion and the Geographies of War," in C. Flint (ed.) *The Geography of War and Peace*, Oxford: Oxford University Press, pp. 144–73.

Provides a framework for identifying and interpreting the role of religion in conflict.

REFERENCES

Afrol.news (2010) "China Woos Taiwan's African Friends." www.afrol.com/articles/22427. Posted July 11, 2010. Accessed May 5, 2011.

Agnew, J. (1987) *Place and Politics*, London: Allen & Unwin.

—— (1994) "The Territorial Trap: The Geographical Assumptions of International Relations Theory," *Review of International Political Economy* 1(1): 53-80.

—— (2002) *Making Political Geography*, London: Arnold.

—— (2003) *Geopolitics: Re-visioning World Politics*, second edition, London: Routledge.

—— (2011) "Waterpower: Politics and the Geography of Water Provision," *Annals of the Association of American Geographers* 101(3): 463–76.

Ahmad, E. (2000) *Confronting Empire*, Cambridge: South End Press.

Ali, M. K. (2005) "Environmental Security of Bangladesh: In the Case of Indo-Bangladesh Relations," *Pakistan Journal of Social Sciences* 3(7): 902–8.

Allen, B. (1996) *Rape Warfare: The Hidden Genocide in Bosnia-Herzegovina and Croatia*, Minneapolis: University of Minnesota Press.

Allen, J. (2003) *Lost Geographies of Power*, Oxford: Blackwell.

Amnesty International (2004) "Israel and the Occupied Territories under the Rubble: House Demolition and Destruction of Land and Property," May 18, 2004. http://web.amnesty.org/library/index/engmde/150332004. Accessed February 5, 2005.

—— (2010) "Amnesty International Public Statement: A Matter of Concern for the Whole of Europe: Human Rights Violations in the North Caucuses," June 15, 2010. www.amnesty.org/en/library/asset/EUR46/020/2010/en/85e68c8f-6f33-4ca9-861e-9ae44867f25c/eur460202010en.pdf. Accessed April 6, 2011.

Amoore, L. (2006) "Biometric Borders: Governing Mobilities in the War on Terror," *Political Geography* 25: 336–51.

—— (2009) "Algorithmic War: Everyday Geographies of the War on Terror," *Antipode* 41: 49–69.

Anderson, M. (1996) *Frontiers: Territory and State Formation in the Modern World*, Oxford: Polity.

Appadurai, A. (1991) "Global Ethnoscapes: Notes and Queries for a Transnational Anthropology," in R. G. Fox (ed.), *Recapturing Anthropology*, Santa Fe: School of American Research Press.

Apple, B. (1998) *School for Rape: The Burmese Military and Sexual Violence*, Washington, DC and Chiang Mai: Earth Rights International.

Arquilla, J. and Ronfeldt, D. (2001) *Networks and Netwars: The Future of Terror, Crime, and Militancy*, Santa Monica, CA: Rand.

Associated Press (2003) "Russian Soldiers' Mothers Fight for Young Draftees," January 14, 2003. www.hrvc.net/news/16b-1-2003.html. Accessed July 14, 2004.

Bacevich, A. J. (2005) *The New American Militarism: How Americans Are Seduced by War*, Oxford and New York: Oxford University Press.

BBC (2004) "Timeline: Chechnya," June 30, 2004. http://news.bbc.co.uk/2/hi/europe/country_profiles/2357267.stm. Accessed July 2, 2004.

—— (2005) "Ex-UN Chief Warns of Water Wars," BBC News, February 2, 2005. http://news.bbc.co.uk/2/hi/africa/4227869.stm. Accessed May 26, 2011.

Beaverstock, J. V., Smith, R. G., and Taylor, P. J. (2000) "World-City Network: A New Metageography?," *Annals of the Association of American Geographers* 90: 123–34.

Bernazzoli, R. and Flint, C. (2009) "From Militarization to Securitization: Finding a Concept that Works," *Political Geography* 28(8): 449–50.

—— (2010) "Embodying the Garrison State? Everyday Geographies of Militarization in American Society," *Political Geography* 29: 157–66.

Bernstein, D. and Kean, L. (1998) "Ethnic Cleansing: Rape as a Weapon of War in Burma," Burma Forum Los Angeles. www.burmaforumla.org/women/rape.htm. Accessed July 18, 2004.

Bigo, D (2001) "The Möbius Ribbon of Internal and External Security(ies)," in M. Albert, D. Jacobson, and Y. Lapid (eds.), *Identities, Borders, Orders: Rethinking International Relations*, Minneapolis: University of Minnesota Press, pp. 91–136.

Billig, M. (1995) *Banal Nationalism*, London: Sage.

Billo, C. G. and Chang, W. (2004) "Cyber Warfare: An Analysis of the Means and Motivations of Selected Nation States," Hanover, NH: Institute for Security Technology Studies at Dartmouth College.

Binford Peay III, J. H. (1995) "The Five Pillars of Peace in the Central Region," *Joint Force Quarterly* 9: 32.

Bix, H. P. (2000) *Hirohito and the Making of Modern Japan*, New York: HarperCollins.

Boserup, E. (1965) *The Conditions of Agricultural Growth: The Economics of Agrarian Change under Population Pressure*, London: Allen & Unwin.

Brogan, P. (1990) *The Fighting Never Stopped*, New York: Vintage Books.

B'TSELEM (2002) "Israel's Policy of House Demolition and Destruction of Agricultural Land in the Gaza Strip." www.btselem.org/English/Publications/Summaries/Policy_of_Destruction.asp. Accessed November 15, 2004.

Byers, M. (2009) "Conflict or Cooperation: What Future for the Arctic?," *Swords and Ploughshares* 17(3): 18–21.

Callahan, M. P. (2004) *Making Enemies: War and State Building in Burma*. Singapore: Singapore University Press.

Calvin, J. B. (1984) "The China India Border War," GlobalSecurity.org. Written 1984. www.globalsecurity.org/military/library/report/1984/CJB.htm. Accessed June 1, 2004.

Castells, M. (1996) *The Rise of the Network Society*. Oxford: Blackwell.

Central Intelligence Agency (2009). Press Release: CIA Opens Center on Climate Change and National Security. www.cia.gov/news-information/press-releases-statements/center-on-climate-change-and-national-security.html. Posted September 25, 2009. Accessed May 22, 2011.

Chase-Dunn, C. and Kaneshiro, M. (2009) "Stability and Change in the Contours of Alliances among Movements in the Social Forum Process," in D. Fasenfest (ed.), *Engaging Social Justice: Critical Studies of 21st Century Social Transformation*, Leiden and Boston: Brill, pp. 119–33.

Chertoff, A. M. (2006) "A Tool We Need to Stop the Next Airliner Plot," *Washington Post*, August 29, A15.

Chhachhi, S. (2002) "Finding Face: Images of Women from the Kashmir Valley," in U. Butalia (ed.), *Speaking Peace: Women's Voices from Kashmir*, London and New York: Zed Books, pp. 189–225.

Chiozza, G. (2002) "Is There a Clash of Civilizations? Evidence from Patterns of International Conflict Involvement 1946–97," *Journal of Peace Research* 39: 711–34.

Clarke, R. (2010) *A Cyber War*. New York: HarperCollins.

Clunan, A. L. and Trinkunas, H. A. (eds.) (2010) *Ungoverned Spaces: Alternatives to State Authority in an Era of Softened Sovereignty*. Stanford, CA: Stanford University Press.

CNN (1998) "Famine May Have Killed 2 Million in North Korea," August 19, 1998. www.cnn.com/WORLD/asiapcf/9808/19/nkorea.famine/. Accessed August 24, 2004.

—— (2004) "Bomb Kills Chechen President," May 9, 2004. www.cnn.com/2004/WORLD/europe/05/09/grozy.blast/index.html. Accessed July 2, 2004.

Cohen, S. (1963) *Geography and Politics in a World Divided*, New York: Random House.

Collins, J. L. (1969) *War in Peacetime*, Boston: Houghton Mifflin.

Cowen, D. and Smith, N. (2009) "After Geopolitics? From the Geopolitical Social to Geoeconomics," *Antipode* 41(1): 22–48.

Cox, K. R. (2002) *Political Geography: Territory, State, and Society*, Oxford: Blackwell.

Cox, K. R., Low, M., and Robinson, J. (2008) *The SAGE Handbook of Political Geography*, London: Sage.

Crenshaw, M. (1981) "Thoughts on Relating Terrorism to Historical Contexts," in M. Crenshaw (ed.), *Terrorism in Context*, University Park: Pennsylvania State University Press, pp. 3–24.

Crossette, B. (1998) "Violation: An Old Scourge of War Becomes Its Latest Crime," *New York Times*, July 14, 1998.

Crutzen, P. (2002) "Geology of Mankind," *Nature* 415: 23.

Cullison, Alan. "Russia's Left-Wing Politicians Retreat from Their Support of U.S.-Led War," *Wall Street Journal* February 5, 2002.

Cumings, B. (1997) *Korea's Place in the Sun*, New York: W. W. Norton & Company.

Dahlman, C. (2005) "Geographies of Genocide and Ethnic Cleansing: The Lessons of Bosnia-Herzegovina," in C. Flint (ed.), *The Geography of War and Peace*, Oxford: Oxford University Press, pp. 174–97.

Dalby, S. (1996) "Reading Robert Kaplan's 'Coming Anarchy'," *Ecumene* 3(4): 472–96.

—— (2003) "Geopolitics, the Bush Doctrine, and War on Iraq," *The Arab World Geographer* 6: 7–18.

—— (2006) "Introduction to Part Four: The Geopolitics of Global Dangers," in G. Ó Tuathail, S. Dalby, and P. Routledge (eds.), *The Geopolitics Reader*, second edition, London and New York: Routledge, pp. 177–87.

—— (2011) *Security and Environmental Change*, Cambridge: Polity Press.

DARPA (2005) "Falcon" Defense Advanced Research Projects Agency. Posted January 7, 2005. www.darpa.mil/tto/programs/falcon.html. Accessed January 31, 2005.

Davis, M. (2001) *Late Victorian Holocausts: El Niño Famines and the Making of the Third World*, London: Verso.

Deudney, D. (1990) "The Case against Linking Environmental Degradation and National Security," *Millennium* 19: 461–76.

—— (1999) "Environmental Security: A Critique," in D. Deudney and R. Matthew (eds.), *Contested Grounds: Security and Conflict in the New Environmental Politics*, Albany: State University of New York Press, pp. 187–219.

Der Derian, J. (2001) *Virtuous War: Mapping the Military–Industrial–Media–Entertainment Network*, Boulder, CO: Westview Press.

Dewan, R. (2002) "What Does Azadi Mean to You?," in U. Butalia (ed.), *Speaking Peace: Women's Voices from Kashmir*, London: Zed Books, pp. 149–61.

Dijkink, G. (1996) *National Identity and Geopolitical Visions: Maps of Pride and Pain*, New York: Routledge.

Dittmer, J. (2010) *Popular Culture, Geopolitics, and Identity*, Lanham, MD: Rowman and Littlefield.

Dodds, K. (2003) "License to Stereotype: James Bond, Popular Geopolitics and the Spectre of Balkanism," *Geopolitics* 8: 125–54.

—— (2009) "From Frozen Desert to Maritime Domain: New Security Challenges in an Ice-Free Arctic," *Swords and Ploughshares* 17(3): 11–14.

—— (2010) "Jason Bourne: Gender, Geopolitics, and Contemporary Representations of National Security," *Journal of Popular Film and Television* 38: 21–33.

Dodds, K. and Atkinson, D. (2000) *Geopolitical Traditions: A Century of Geopolitical Thought*, London: Routledge.

Donnan, H. and Wilson, T. M. (1999) *Borders: Frontiers of Identity, Nation and State*, Oxford: Berg.

Dowler, L. (2005) "Amazonian Landscapes: Gender, War, and Historical Repetition," in C. Flint (ed.), *The Geography of War and Peace*, Oxford: Oxford University Press, pp. 133–48.

Dowler, L. and Sharp, J. (2001) "A Feminist GeoPolitics," *Space and Polity* 5(3): 165–76.

Eckert, C. J., Lee, K.-B., Lew, Y., Robinson, M., and Wagner, E. W. (1990) *Korea Old and New*, Seoul: Ilchokak Publishers.

The Economist (2004) "Plots, Alarms, and Arrests," August 14, 2004, pp. 22–4.

Eksteins, M. (1989) *Rites of Spring: The Great War and the Birth of the Modern Age*, New York: Anchor Books.

Elden, S. (2009) *Terror and Territory: The Spatial Extent of Sovereignty.* Minneapolis: University of Minnesota Press.

Enloe, C. (1983) *Does Khaki Become You? The Militarisation of Women's Lives*, London: Pluto Press.

—— (1990) *Bananas, Beaches, and Bases: Making Feminist Sense of International Politics*, Berkeley: University of California Press.

—— (2004) *The Curious Feminist: Searching for Women in a New Age of Empire*, Berkeley: University of California Press.

Etzold, T. H. and Gaddis, J. L. (1978) *Containment: Documents on American Policy and Strategy, 1945–1950.* New York: Columbia University Press.

Evangelista, M. (2002) *The Chechen Wars: Will Russia Go the Way of the Soviet Union?*, Washington, DC: Brookings Institution Press.

Falah, G.-W. (2005) "Peace, Deception, and Justification for Territorial Claims: The Case of Israel," in C. Flint (ed.), *The Geography of War and Peace*, Oxford: Oxford University Press, pp. 297–320.

Falah, G.-W. and Flint, C. (2004) "Geopolitical Spaces: The Dialectic of Public and Private Space in the Palestine–Israel Conflict," *The Arab World Geographer* 7: 117–34.

Flint, C. (2001) "The Geopolitics of Laughter and Forgetting: A World-Systems Interpretation of the Post-Modern Geopolitical Condition," *Geopolitics* 6: 1–16.

—— (20003a) "Terrorism and Counterterrorism: Geographic Research Questions and Agendas," *Professional Geographer* 55: 161–9.

—— (2003b) "Geographies of Inclusion/Exclusion," in S. L. Cutter, D. B. Richardson, and T. J. Wilbanks (eds.), *The Geographical Dimensions of Terrorism*, New York: Routledge, pp. 53–8.

—— (2004) "The 'War on Terrorism' and the 'Hegemonic Dilemma': Extraterritoriality, Reterritorialization, and the Implications for Globalization," in J. O'Loughlin, L. Staeheli, and E. Greenberg (eds.), *Globalization and Its Outcomes*, New York: The Guilford Press, pp. 361–85.

—— (2005) "Dynamic Metageographies of Terrorism: The Spatial Challenges of Religious Terrorism and the 'War on Terrorism'," in C. Flint (ed.), *The Geography of War and Peace*, Oxford: Oxford University Press, pp. 198–216.

Flint, C., Adduci, M., Chen, M., and Chi, S.-H. (2009) "Mapping the Dynamism of the United States' Geopolitical Code: The Geography of the State of the Union Speeches, 1998–2008," *Geopolitics* 14: 604–29.

Flint, C. and Taylor, P. J. (2011) *Political Geography: World-economy, Nation-state, and Locality*, sixth edition, Harlow: Pearson Education.

The Fog of War: An Errol Morris Film (2003), Sony Pictures Classic, Inc.

Food and Agriculture Organization (2004) *The State of Food and Agriculture 2003–2004*, Rome: Food and Agriculture Organization of the United Nations.

Friedman, T. L. (1995) *From Beirut to Jerusalem*, New York: Anchor Books.

Fussell, P. (1990) *Wartime: Understanding and Behavior in the Second World War*, Oxford: Oxford University Press.

Galtung, J. (1964) "What Is Peace Research?," *Journal of Peace Research* 1(1): 1–4.

—— (1996) *Peace by Peaceful Means: Peace and Conflict, Development and Civilization*, London, Thousand Oaks, CA and Delhi: Sage.

Geren, P. and Casey Jr., General G. W. (2009) "A Statement on the Posture of the United States Army 2009," Washington, DC: United States Senate and House of Representatives.

German, T. C. (2003) *Russia's Chechen War*, London: Routledge.

Giles, W. and Hyndman, J. (2004) *Sites of Violence: Gender and Conflict Zones*, Berkeley: University of California Press.

Gilmartin, M. and Kofman, E. (2004) "Critically Feminist Geopolitics," in L. A. Staeheli, E. Kofman, and L. J. Peake (eds.), *Mapping Women, Making Politics*, New York and London: Routledge, pp. 113–25.

Glassner, M. and Fahrer, C. (2004) *Political Geography*, third edition, Hoboken, NJ: Wiley and Sons.

GOM, Government of Myanmar (1994) *Our Three Main National Causes*, Yangon: News and Periodicals Enterprise, Ministry of Information.

Gottman, J. (1973) *The Significance of Territory*, Charlottesville: University Press of Virginia.

Gramsci, A. (1971) *Selections from Prison Notebooks*, London: Lawrence and Wishart.

Grundy-Warr, C. and Dean, K. (2011) "Not Peace, Not War: The Myriad Spaces of Sovereignty, Peace and Conflict in Myanmar/Burma," in S. Kirsch and C. Flint (eds.), *Reconstructing Conflict: Integrating War and Post-War Geographies*. Farnham: Ashgate, pp. 91–114.

Gurr, T. R. (2000) *Peoples against States: Minorities at Risk in the New Century*, Washington, DC: United States Institute of Peace Press.

Gusev, D. "Vladimir Zhirinovsky," October 17, 1996. www.cs.indiana.edu/~dmiguse/Russian/vzbio.html. Accessed July 29, 2004.

Haggett, P. (1979) *Geography: A Modern Synthesis*, third edition, New York: Harper and Row.

Haraway, D. (1998) "Situated Knowledges: The Science Question in Feminism and the Privilege of Partial Perspective," *Feminist Studies* 14: 575–99.

Harkavy, R. E. (2007) *Strategic Basing and the Great Powers, 1200–2000*, New York: Routledge.

Harris, L. M. (2005) "Navigating Uncertain Waters: Geographies of Water and Conflict, Shifting Terms and Debates," in C. Flint (ed.), *The Geography of War and Peace*, Oxford: Oxford University Press, pp. 259–79.

Harvey, D. (2003) *The New Imperialism*, Oxford: Oxford University Press.

Hass, A. (2000) *Drinking the Sea at Gaza*, New York: Owl Books.

Haushofer, K., Obst, E., Lautensach, H., and Maull, O. (1928) *Bausteine zur Geopolitik*, Berlin: Kurt Vowinckel Verlag.

Hedges, C. (2003) *War Is a Force That Gives Us Meaning*, New York: Anchor Books.

Henrikson, A. K. (2005) "The Geography of Diplomacy," in C. Flint (ed.), *The Geography of War and Peace*, Oxford: Oxford University Press, pp. 369–94.

Heo, U. and Hyun, C.-M. (2003) "The 'Sunshine' Policy Revisited," in U. Heo and S. A. Horowitz (eds.), *Conflict in Asia*, Westport, CT: Praeger, pp. 89–103.

Herb, G .H. and Kaplan, D. H. (1999) *Nested Identities*, Lanham, MD: Rowman and Littlefield.

Herbst, J. I. (2000) *States and Power in Africa: Comparative Lessons in Authority and Control*, Princeton, NJ: Princeton University Press.

Herod, A. and Wright, M. W. (2002) *Geographies of Power: Placing Scale*, Oxford: Blackwell.

Hoare, J. and Pares, S. (1999) *Conflict in Korea*, Denver, CO: ABC-CLIO.

Hoffman, B. (1998) *Inside Terrorism*, New York: Columbia University Press.

—— (2002) "Lessons of 9/11," Joint Inquiry Staff Request, 8 October 2002.

Holworth, J. (2011) "The European Union in (In-)Action: Brussels and the Arab Spring," *Swords and Ploughshares* 19(1): forthcoming.

Hubbard, P., Kitchin, R., Bartley, B., and Fuller, D. (2002) *Thinking Geographically: Space, Theory and Contemporary Human Geography*, London: Continuum.

Human Rights Watch (1999) "The Aftermath: The Ongoing Issues Facing Kosovar Albanian Women." http://hrw.org/reports/2000/fry/Kosov003.htm#P38_1195. Accessed July 22, 2004.

—— (2004) "On the Situation of Ethnic Chechens in Moscow," February 24, 2004. www.hrw.org/backgrounder/eca/russia032003.htm. Accessed July 7, 2004.

Huntington, S. (1993) "The Clash of Civilizations," *Foreign Affairs* 72: 22–49.

Hyndman, J. (2003) "Aid, Conflict, and Migration: The Canada–Sri Lanka Connection," *Canadian Geographer* 47(3): 251–68.

—— (2004) "Mind the Gap: Bridging Feminist and Political Geography through Geopolitics," *Political Geography* 23: 307–22.

Indian Express Group (2001) "The Four Indo-Pak Wars." www.expressindia.com/kashmir/kashmirlive/wars.html. Accessed June 10, 2004.

Islamic Republic of Pakistan (2004) "Frequently Asked Questions," May 25, 2004. www.infopak.gov.pk/public/kashmir/q&a.htm#didnot. Accessed June 2, 2004.

Jiang, J. and Sinton, J. (2011) *Overseas Investments by Chinese National Oil Companies: Assessing the Drivers and Impacts*, Paris: International Energy Agency.

Johnson, S. E. and Long, D. (eds.) (2007) *Coping with the Dragon: Essays on the PLA Transformation and the U.S. Military*, Washington, DC: Center for Technology and National Security Policy.

Johnston, R. J. and Sidaway, J. D. (2004) *Geography and Geographers: Anglo-American Human Geography since 1945*, sixth edition, London: Arnold.

Juergensmeyer, M. (2000) *Terror in the Mind of God*, Berkeley: University of California Press.

Kaplan, R. (1994) "The Coming Anarchy," *Atlantic Monthly* 273: 44–76.

Kapuscinski, R. (1992) *The Soccer War*, New York: Vintage Books.

Kearns, G. (2009) *Geopolitics and Empire: The Legacy of Halford Mackinder*, Oxford and New York: Oxford University Press.

Khashan, H. (2000) *Arabs at the Crossroads: Political Identity and Nationalism*, Gainesville: University of Florida Press.

Kirsch, S. and Flint, C. (2011a) "Introduction: Reconstruction and the Worlds that War Makes," in S. Kirsch and C. Flint (eds.), *Reconstructing Conflict: Integrating War and Post-War Geographies*, Farnham: Ashgate, pp. 3–28.

—— (eds.) (2011b) *Reconstructing Conflict: Integrating War and Post-War Geographies*, Farnham: Ashgate.

Klare, M. T. (1996) *Rogue States and Nuclear Outlaws: America's Search for a New Foreign Policy*, New York: Hill and Wang.

—— (2009) *Rising Powers, Shrinking Planet: The New Geopolitics of Energy*, New York: Holt Paperbacks.

Knezys, S. and Sedlickas, R. (1999) *The War in Chechnya*, College Station: Texas A&M University Press.

Knox, P. L. and Marston, S (1998) *Places and Regions in Global Context: Human Geography*, Upper Saddle River, NJ: Prentice-Hall.

Konrád, G. (1984) *Antipolitics*, San Diego: Harcourt, Brace, and Jovanovich.

Koopman, S. (forthcoming) "Altergeopolitics: Other Securities Are Happening," *Geoforum*.

Kriesberg, L. (1997) "Social Movements and Global Transformation," in J. Smith, C. Chatfield, and R. Pagnucco (eds.), *Transnational Social Movements and Global Politics*, Syracuse, NY: Syracuse University Press, pp. 3–18.

Kuus, M. and Agnew, J. A. (2008) "Theorizing the State Geographically: Sovereignty, Subjectivity, Territoriality," in K. R. Cox, M. Low, and J. Robinson (eds.), *The SAGE Handbook of Political Geography*, Los Angeles, London, New Delhi, and Singapore: Sage.

LaFraniere, S. (2003) "Rivalry Fragments Russia's Liberal," *Washington Post* July 2, 2003. www.ncsj.org/AuxPages/020703WPost.shtml. Accessed July 27, 2004.

Lambrecht, C. T. (2004) "Oxymoronic Development: The Military as Benefactor in the Border Regions of Burma," in C. R. Duncan (ed.), *Civilizing the Margins: Southeast Asian Government Policies for the Development of Minorities*, New York: Cornell University Press, pp. 150–81.

Laqueur, W. (1987) *The Age of Terrorism*, Boston: Little, Brown, and Company.

Le Billon, P. (2005) "The Geography of 'Resource Wars'," in C. Flint (ed.), *The Geography of War and Peace*, Oxford and New York: Oxford University Press, pp. 217–41.

Lenin, V. I. (1939) *Imperialism: The Highest Stage of Capitalism*, New York: International Publishers.

Levering, R. B. (1997) "Brokering the Law of the Sea Treaty: The Neptune Group," in J. Smith, C. Chatfield, and R. Pagnucco (eds.), *Transnational Social Movements and Global Politics*, Syracuse, NY: Syracuse University Press, pp. 225–39.

Lewis, B. (2002) *What Went Wrong? Western Impact and Middle Eastern Response*, Oxford: Oxford University Press.

Lewis, M. W. and Wigen, K. E. (1997) *The Myth of Continents: A Critique of Metageography*, Berkeley: University of California Press.

Lineback, N. G. (2002) "India's Religious Quarrels," Appalachian State University. www.geo. appstate.edu/616_032202indiac.pdf. Accessed June 9, 2004.

Lintner, B. (1990) *Outrage: Burma's Struggle for Democracy*. Bangkok: White Lotus.

Long, J. M. (2004) *Saddam's War of Words*, Austin: University of Texas Press.

Lundestad, I. (2009) "US Security Policy and Regional Relations in a Warming Arctic," *Swords and Ploughshares* 17(3): 15–17.

Lynn, III, W. J. (2010) "Defending a New Domain: The Pentagon's Cyberstrategy," *Foreign Affairs* September/October 2010, pp. 97–108.

McAlister, M. (2001) *Epic Encounters: Culture, Media, and US Interests in the Middle East, 1945–2000*, Berkeley: University of California Press.

Maclean, W. (2010) "Iran 'First Victim of Cyberwar'," *The Scotsman* September 25, 2010. http://news.scotsman.com/world/Iran-39first-victim-of-cyberwar39.6550278.jp. Accessed April 25, 2011.

Malik, I. (2002) *Kashmir: Ethnic Conflict International Dispute*, Oxford: Oxford University Press.

Malthus, T. (1970/1798) *An Essay on the Principle of Population*, Harmondsworth: Penguin.

Mamadouh, V. (2005) "Geography and War, Geographers and Peace," in C. Flint (ed.), *The Geography of War and Peace*, Oxford and New York: Oxford University Press, pp. 26–60.

Mankoff, J. (2010) *The Russian Economic Crisis*, Washington, DC: Council on Foreign Relations.

Mann, M. (1986) *The Sources of Social Power*, Vol. 1, New York: Cambridge University Press.

Mansfield, P. (1992) *The Arabs*, third edition, New York: Penguin.

Martínez, O. J. (1994) *Border People: Life and Society in the U.S.–Mexico Borderlands*, Tucson: University of Arizona Press.

Massey, D. (1994) *Space, Place, and Gender*, Minneapolis: University of Minnesota Press.

Meadows, D. H., Meadows, D. L., Randers, J., and Behrens III, W. W. (1974) *The Limits to Growth*, London: Pan.

Megoran, N. (2010) "Towards a Geography of Peace: Pacific Geopolitics and Evangelical Christian Crusade Apologies," *Transactions, Institute of British Geographers* NS 35: 382–98.

—— (2011) "War *and* Peace? An Agenda for Research and Practice in Geography," *Political Geography* 1-12.

Mercille, J. (2008) "The Radical Geopolitics of US Foreign Policy: Geopolitical and Geoeconomic Logics of Power," *Political Geography* 27(5): 570–86.

Modelski, G. (1987) *Long Cycles in World Politics*, Seattle: University of Washington Press.

Modelski, G. and Thompson, W. (1995) *Leading Sectors and World Powers*, Columbia: University of South Carolina Press.

Moore, T. G. (2005) "Chinese Foreign Policy in the Age of Globalization," in Y. Deng and F.-L. Wang (eds.), *China Rising: Power and Motivation in Chinese Foreign Policy*, Lanham, MA: Rowman and Littlefield, pp. 121–58.

Moran, D. (2011) *Climate Change and National Security: A Country-Level Analysis*, Washington, DC: Georgetown University Press.

Morrissey, J. (2008). "The Geoeconomic Pivot of the Global War on Terror: US Central Command and the War in Iraq," in D. Ryan and P. Kiely (eds.), *America and Iraq: Policy-Making, Intervention and Regional Politics since 1958*, New York: Routledge, pp. 103–22.

Mosse, G. L. (1975) *The Nationalization of the Masses*, New York: H. Fertig.

Murphy, A. B. (2005) "Territorial Ideology and Interstate Conflict: Comparative Considerations," in C. Flint (ed.), *The Geography of War and Peace*, Oxford: Oxford University Press, pp. 280–96.

Narangoa, L. (2004) "Japanese Geopolitics and the Mongol Lands, 1915–1945," *European Journal of East Asian Studies* 3(1): 45–67.

National Geospatial-Intelligence Agency (2004) *Geospatial Intelligence (GEOINT) Basic Doctrine*, Washington, DC: National Geospatial-Intelligence Agency.

National Security Strategy of the United States of America (2002) Washington, DC: The White House.

Netanyahu, B. (2000) *A Durable Peace: Israel and its Place among the Nations*, New York: Warner Books.

Newman, D. (2005) "Conflict at the Interface: The Impact of Boundaries and Borders on Contemporary Ethnonational Conflict," in C. Flint (ed.), *The Geography of War and Peace*, Oxford: Oxford University Press, pp. 321–44.

NSC-68: United States Objectives and Programs for National Security (1950). April 7, 1950. www.fas.org/irp/offdocs/nsc-hst/nsc-68.htm. Accessed July 22, 2004.

O'Lear, S. (2004) "Resources and Conflict in the Caspian Sea," *Geopolitics* 9: 161–86.

O'Loughlin, J. (1994) *Dictionary of Geopolitics*, Westport, CT: Greenwood Press.

O'Loughlin, J. and Raleigh, C. (2008) "Spatial Analysis of Civil War Violence," in K. Cox, M. Low, and J. Robinson (eds.), *The SAGE Handbook of Political Geography*, London: Sage, pp. 493–508.

Ó Tuathail, G. (1996) *Critical Geopolitics*, Minneapolis: University of Minnesota Press.

—— (2006) "General Introduction: Thinking Critically about Geopolitics," in G. Ó Tuathail, S. Dalby, and P. Routledge (eds.), *The Geopolitics Reader*, London: Routledge, pp. 1–14.

Oas, I. (2005) "Shifting the Iron Curtain of Kantian Peace: NATO Expansion and the Modern Magyars," in C. Flint (ed.), *The Geography of War and Peace*, Oxford: Oxford University Press, pp. 395–414.

Oberdorfer, D. (2001) *The Two Koreas.* Indianapolis: Basic Books.

Painter, J. (1995) *Politics, Geography and "Political Geography"*, London: Arnold.

Parker, G. (1985) *Western Geopolitical Thought in the Twentieth Century*, London: Croom Helm.

Patrick, S. (2007) "'Failed' States and Global Security: Empirical Questions and Policy Dilemmas," *International Studies Review* 9: 644–62.

Peet, R. (1998) *Modern Geographical Thought*, Oxford: Blackwell.

Peluso, N. L. and Vandergeest, P. (2011) "Political Ecologies of War and Forests: Counterinsurgencies and the Making of National Natures," *Annals of the Association of American Geographers* 101(3): 587–608.

Podur, J. (2002) "Kashmir Timeline," January 10, 2002. www.zmag.org/southasia/kashtime.htm. Accessed November 1, 2004.

Polgreen, L. (2005) "Darfur's Babies of Rape Are on Trial from Birth," *New York Times*, February 11, p. A1.

Politkovskaya, A. (2003) *A Small Corner of Hell: Dispatches from Chechnya*, Chicago: University of Chicago Press.

Prescott, J. R. V. (1987) *Political Frontiers and Boundaries*, London: Unwin Hyman.

Priest, D. (2003) *The Mission: Waging War and Keeping Peace with America's Military*, New York: W. W. Norton & Co.

Raina, S. B. (2002) "Leaving Home," in U. Butalia (ed.), *Speaking Peace: Women's Voices from Kashmir*, London: Zed Books, pp. 178–84.

Raleigh, C. and Urdal, H. (2007) "Change, Environmental Degradation and Armed Conflict," *Political Geography* 26(6): 674–94.

Raman, B. (2001) "Pakistan's Inter-Service Intelligence," South Asia Analysis Group. August 1, 2001. www.saag.org/papers3/paper287.html. Accessed May 25, 2005.

Ramet, S. P. (1999) *Gender Politics in the Western Balkans*, University Park: Pennsylvania State University Press.

Ranstorp, M. (1998) "Interpreting the Broader Context and Meaning of bin Laden's *fatwa*," *Studies in Conflict and Terrorism* 21: 321–30.

Rapoport, D. C. (2001) "The Fourth Wave: September 11 in the History of Terrorism," *Current History* 100: 419–24.

Rose, G. (1997) "Situating Knowledges: Positionality, Reflexivities and Other Tactics," *Progress in Human Geography* 21: 305–20.

Sack, R. (1986) *Human Territoriality: Its Theory and History.* Cambridge: Cambridge University Press.

Said, E. (1979) *Orientalism*, New York: Vintage Books.

—— (2004) *From Oslo to Iraq and the Roadmap*, London: Bloomsbury.

Salmon, T. C. and Shepherd, A. (2003) *Toward a European Army: A Military Power in the Making*? Boulder, CO: Lynne Rienner.

Sankalp, S. (2003) "Spotlight," Indian National Congress. July 9, 2003. www.indiannational congress.org/shimla_sankalp.shtml. Accessed May 30, 2004.

Santhanam, K., Sreedhar, Saxena, S., and Manish (2003) *Jihadis in Jammu and Kashmir*, New Delhi: Sage.

Schmid, A. P., Jongman, A. J., and Horowitz, I. (1988) *Political Terrorism: a new guide to actors, authors, concepts, data bases, theories, and literature.* Amsterdam: Transaction Books.

Schofield, C., Newman, D., Drysdale, A., and Brown, J. A. (eds.) (2002) *The Razor's Edge: International Boundaries and Political Geography*, London: Kluwer Law International.

Schofield, R. (2003) "Britain and Kuwait's Borders, 1902–23," in B. J. Slot (ed.), *Kuwait: The Growth of a Historic Identity*, London: Arabian Publishing, pp. 58–94.

Schwarzkopf, General H. N. (1990) Statement before the Senate Armed Services Committee on the Posture of United States Central Command. Washington, DC: Senate Armed Forces Committee, February 8, 1990.

Selth, A. (1996) *Transforming the Tatmadaw: The Burmese Armed Forces since 1988*, Canberra: Strategic and Defence Studies Centre, Australian National University.

—— (2001) *Burma: A Strategic Perspective*, San Francisco, CA: Asia Foundation.

Shapiro, M. (2007) "The New Violent Cartography," *Security Dialogue* 38: 291–313.

Sharp, J. (2000a) *Condensing the Cold War:* Reader's Digest *and American Identity*, Minneapolis: University of Minnesota Press.

—— (2000b) "Refiguring Geopolitics: The *Reader's Digest* and Popular Geographies of Danger at the End of the Cold War," in K. Dodds and D. Atkinson (eds.), *Geopolitical Traditions: A Century of Geopolitical Thought*, London and New York: Routledge, pp. 332–52.

Shlaim, A. (1995) *War and Peace in the Middle East*, revised and updated, New York: Penguin Books.

—— (2001) *The Iron Wall: Israel and the Arab World*, New York: W. W. Norton and Co.

Siow, M. W.-S. (2010) "Chinese Domestic Debates on Soft Power and Public Diplomacy," *Asia Pacific Bulletin* 86, December 7, 2010. Washington, DC: East-West Center.

Slot, B. J. (2003) "Kuwait: The Growth of a Historic Identity," in B. J. Slot (ed.) *Kuwait: The Growth of a Historic Identity*, London: Arabian Publishing, pp. 5–29.

Smith, A. D. (1993) *National Identity*, Reno: University of Nevada Press.

—— (1995) *Nations and Nationalism in a Global Era*, Cambridge: Polity Press.

Smith, J. (1997) "Characteristics of the Modern Transnational Social Movement Sector," in J. Smith, C. Chatfield, and R. Pagnucco (eds.), *Transnational Social Movements and Global Politics*, Syracuse, NY: Syracuse University Press, pp. 42–58.

Smith, N. (2003) *American Empire: Roosevelt's Geographer and the Prelude to Globalization*, Berkeley: University of California Press.

Sorokin, P. A. (1937) *Social and Cultural Dynamics*, Volume 3, New York: American Book Company.

Srivastava, M. (2001) *International Dimensions of Ethnic Conflict: A Case Study of Kashmir and Northern Ireland*, New Delhi: Bhavana Books and Print.

Staeheli, L. A. and Kofman, E. (2004) "Mapping Gender, Making Politics: Toward Feminist Political Geographies," in L. A. Staeheli, E. Kofman, and L. J. Peake (eds.), *Mapping Women, Making Politics*, New York and London: Routledge, pp. 1–13.

Staeheli, L. A., Kofman, E., and Peake, L. J. (eds.) (2004) *Mapping Women, Making Politics*, New York and London: Routledge.

State of the Union Speech (2002). January 29, 2002. www.whitehouse.gov/news/releases/2002/01/print/20020129-11.html. Accessed July 26, 2004.

Steinberg, D. (1984) "Constitutional and Political Bases of Minority Insurrections in Burma," in J.-J. Lim and S. Vani (eds.), *Armed Separatism in Southeast Asia*, Singapore: Institute of Southeast Asian Studies.

—— (2007) "Legitimacy in Burma/Myanmar: Concepts and Implications," in N. Ganesan and K. Y. Hlaing (eds.), *Myanmar: State, Society and Ethnicity*, Singapore: Institute of Southeast Asian Studies, pp. 109–42.

Stump, R. W. (2005) "Religion and the Geographies of War," in C. Flint (ed.), *The Geography of War and Peace*, Oxford: Oxford University Press, pp. 144–73.

Sumartojo, R. (2004) "Contesting Place: Antigay and -Lesbian Hate Crime in Columbus, Ohio," in C. Flint (ed.), *Spaces of Hate*, New York and London: Routledge, pp. 87–107.

Taylor, P. J. (1990) *Britain and the Cold War: 1945 as a Geopolitical Transition*, London: Pinter.

Taylor, P. J. and Flint, C. (2000) *Political Geography: World-Economy, Nation-State, and Locality*, fourth edition, Harlow: Prentice-Hall.

Thompson, E. P. (1985) *The Heavy Dancers*, London: Merlin Press.

Tuchman, B. (1962) *The Guns of August*, New York: Dell.

United Nations (1982) "Statement by the President of Law of the Sea Conference at Opening Meeting of Montego Bay Session," United Nations Press Release, SEA/MP/Rev.1 (6 Dec.), pp. 3, 4–5.

Upadhyay, R. (2000) "The Muslim Agenda of the BJP – Will It Work?" South Asia Analysis Group. September 28, 2000. www.saag.org/papers2/paper149.html. Accessed June 4, 2004.

Vaicikonas, J. (2011) *Strange Bedfellows: North Korean WMD Trading Relationships*, Washington, DC: The Fund for Peace.

Wallerstein, I. (1979) *The Capitalist World-Economy*, Cambridge: Cambridge University Press.

—— (1984) *The Politics of the World-Economy*, Cambridge: Cambridge University Press.

Walsh, N. P. "'A Second Chechnya.' Ingushetia Dispatch," June 18, 2004. www.guardian. co.uk/elsewhere/journalist/story/0,7792,1242054,00.html. Accessed July 14, 2004.

Weizman, E. (2007) *Hollow Land: Israel's Architecture of Occupation*, London: Verso Press.

Williams. P. (2007) "Hindu–Muslim Brotherhood: Exploring the Dynamics of Communal Relations in Varanasi, North India," *Journal of South Asian Development* 2(2): 153–76.

Williams, P. and McConnell, F. (forthcoming) "Critical Geographies of Peace," *Antipode*.

Women's Initiative (2002) "Women's Testimonies from Kashmir," in U. Butalia (ed.), *Speaking Peace: Women's Voices from Kashmir*, London: Zed Books, pp. 82–95.

Women's Organizations from Burma, Women's Affairs Department and National Coalition Government of the Union of Burma (2000) "Burma: The Current State of Women in Conflict Areas." January 2000. www.earthrights.org/women/ShadowWRP.doc. Accessed July 20, 2004.

Woodward, R. (2004) *Military Geographies*, Malden, MA: Blackwell.

World Bank (2011) *World Development Report 2011*, Washington, DC: World Bank.

World Oil (2003) "Greatest Oil Reserves by Country," *World Oil* 224.

INDEX